Passing the Mathematics
Test for Elementary Teachers

Passing the Mathematics Test for Elementary Teachers

Offering a Pathway to Success

Margie Pearse
Diane Devanney
Darla Nagy

ROWMAN & LITTLEFIELD
Lanham • Boulder • New York • London

Published by Rowman & Littlefield
A wholly owned subsidiary of The Rowman & Littlefield Publishing Group, Inc.
4501 Forbes Boulevard, Suite 200, Lanham, Maryland 20706
www.rowman.com

Unit A, Whitacre Mews, 26-34 Stannary Street, London SE11 4AB

Copyright © 2015 by Margie Pearse, Diane Devanney, and Darla Nagy

All rights reserved. No part of this book may be reproduced in any form or by any electronic or mechanical means, including information storage and retrieval systems, without written permission from the publisher, except by a reviewer who may quote passages in a review.

British Library Cataloguing in Publication Information Available

Library of Congress Cataloging-in-Publication Data
Pearse, Margie, author.
  Passing the mathematics test for elementary teachers : offering a pathway to success / Margie Pearse, Diane Devanney, and Darla Nagy.
     pages cm
  Includes bibliographical references and index.
  ISBN 978-1-4758-1083-7 (cloth : alk. paper) — ISBN 978-1-4758-1084-4 (pbk. : alk. paper) — ISBN (invalid) 978-1-4758-1085-1 (electronic)
  1. Mathematics—Vocational guidance. 2. Elementary school teachers—Certification. 3. Mathematics teachers—Certification. I. Devanney, Diane, 1955- author. II. Nagy, Darla, 1957- author. III. Title.
  QA10.5.P43 2015
  510.76—dc23
                                                                                   2014033642

∞^{TM} The paper used in this publication meets the minimum requirements of American National Standard for Information Sciences—Permanence of Paper for Printed Library Materials, ANSI/NISO Z39.48-1992.

Printed in the United States of America

Contents

Acknowledgments		vii
1	How to Pass the Test!	1
	Preparing for the Test: Proven Strategies for Successful Test Preparation	1
	Taking the Test: Proven Strategies for Successful Test-Taking	4
2	Understanding Numbers and the Number System	9
	Number Systems: The Hierarchy of Numbers	9
	Decimals	12
	Scientific Notation	13
	Percents	15
	Simple Interest and Compound Interest	20
	Exponential Notation	24
3	Pre-Algebra	27
	Analyzing and Extending a Variety of Patterns	27
	Solving One-Variable Equations and Inequalities Algebraically	31
	Using the Concepts of Variable, Equality, and Equation to Generate, Interpret, and Evaluate Algebraic Expressions	37
	Manipulating Algebraic Expressions and Solving Equations Using a Variety of Techniques	40
	Simplifying or Combining Like Terms	41
	Basic Distributive Property	42
	Factoring an Expression	43
	Applying Algebraic Principles to Represent and Solve Word Problems Involving Ratios and Proportions	48
4	Algebra	53
	Graphing Lines and Inequalities	53
	Graphing Parallel and Perpendicular Lines	59
	Solving Systems of Linear Equations	62

Solving and Graphing Linear Inequalities and Systems of Linear Inequalities	68
Nonlinear Functions and Real-World Situations	73

5 Measurement and Geometry — 79

Measurement: The U.S. Customary System	79
Measurement: The Metric System	81
Perimeter, Area, Volume, and Surface Area	83
Circles	84
Triangles	86
Quadrilaterals	92
Irregular Figures/Shaded Regions	94
Sphere	97
Angles and Lines	99
Coordinate Geometry	102
Transformations of Points and Geometric Figures	104

6 Statistics and Probability — 109

Measures of Central Tendency and Spread	109
Displaying Data and Statistical Information	112
Analyzing and Drawing Inferences from Data Presented in Different Formats	118
Probability: Simple, Compound, Independent, Dependent, Conditional	121
Counting Principles: Permutations and Combinations	130

Appendix A: Full Practice Test 1	135
Appendix B: Full Practice Test 2	159
Appendix C: Full Practice Test 3	181
Index	205

Acknowledgments

We are especially grateful to the following people, who recognized the need for this book and immediately supported our efforts in creating it. To our publisher and vice president of Rowman and Littlefield Education, Tom Koerner, you are a true visionary. Your fervor to provide what is best for education inspired us to deliver and deliver big. To Carlie Wall, our associate editor, you are a treasure! Your hard work and dedication are much appreciated. Your willingness to answer any and all questions along the way often gave us the *oomph* we needed to keep pressing on. To Christopher Basso, thank you for all your hard work and patience throughout the editing process.

To Bridget Collington, thank you for skillfully creating every figure included in this book. Your diligence and patience are to be commended. Any school district would be fortunate to hire such a dedicated and energetic math teacher.

A big thank you goes out to the math department at Cabrini College. Your ideas, support, and wisdom are clearly evident in every page. To the education students at Neumann University and Cabrini College, thank you for testing out every problem in this book. Your mission to do something extraordinary will come to fruition when hundreds of pre-service teachers across the country are passing as a result of this book.

—Margie, Diane, and Darla

I am eternally grateful for my Mom, Gloria DiJoseph. Her spunk, her drive, and her zest for life became the inspiration for this work. Thank you to my best friend and soul mate, Chris Pearse. You recognize things in me that I have never imagined for myself, and that vision motivated me to press on once again in my writing. To my three children, Chris Pearse, Drew Pearse, and Gloria Pearse: your passion for life is contagious. Thank you for sharing life's joys with me. To my favorite little guy in the whole wide world, Ayden Flood: you fill my heart with smiles. Thank you to my siblings, Linda Harris, Joanne Howarth, Nancy Gauzza, and Joe Charley. You never doubt me, even when I doubt myself. That confidence pushed me through to the end. To my lifelong friend, Maureen Koob: your fifty-year friendship has often been the wind beneath my wings. To my dear friends, Jennifer Ennis, Tina Quiram, and Christy Rydel: your continued friendship and support are treasured. Last but certainly not least, thank you to Diane and Darla for living out this dream with me. You are both gems. Your integrity, hard work, and brilliance made this experience a blast!

—Margie Pearse

I dedicate this book to those people who have made an impact in my life. To my late father, Neil Summers, whose math games on long car trips began my lifelong love of learning. To my mother, Marie Summers, who continues to inspire me with her spunk and wit. To my husband, Joseph Devanney, who is my best friend and my soul mate. My heart still skips a beat when you walk into a room. To our daughter, Katie Devanney, who stole my heart the first time I held her. I love you more than you'll ever know. To my colleagues Margie and Darla, who teach me something new every day. You make me look forward to coming to work! And, to the Sisters, Servants of the Immaculate Heart of Mary, for guiding me through grade school and high school at Villa Maria Academy, Malvern, PA, and college at Immaculata University, Immaculata, PA. Your love remains with me today.

—Diane Devanney

Thank you to my family and friends for their help in writing this book. To my wonderful kids, Andy and Jason, who have encouraged me on my journey deciding "What I want to do when I grow up!" To my sisters, Wendy and Connie, who keep me motivated when I get overwhelmed. To my mom, Darleen, who is always proud of me, no matter what I am doing! To my wonderful friend Cathy, who put up with me and provided needed distractions as I worked on this book. To my coauthors, Diane and Margie, it is a pleasure working with you at Cabrini College. And finally, to my dear husband, Pete, whose gentle "Why do you want to teach?" wakes me up and reminds me of my love of math and my desire to help our future teachers embrace, not fear, math and to learn how to unlock the mysteries of math for the next generation. I love you all!

—Darla Nagy

Chapter One

How to Pass the Test!

Dear Future Teachers:

In this chapter, you will find proven methods for both preparing for and taking standardized tests. As mathematics and education professors, we can attest to the effectiveness of each strategy offered. Every idea and each lesson provided are integral to the pre-service mathematics test-prep courses we teach and students are seeing amazing results. They are passing! We are thrilled to share these discoveries with you and wish you all the best in your future career in education.

PREPARING FOR THE TEST: PROVEN STRATEGIES FOR SUCCESSFUL TEST PREPARATION

It is important to follow a long-term study plan in order to successfully prepare for any standardized assessment. Ideally, an eight-week study schedule is perfect. But if that is impossible, a condensed version can work in four weeks.

Step 1: Register and Pay for the Test

Go to your state's Educator Certification Tests website, register, and pay for the pre-service Mathematics module. If possible, select a date that is at least two months from the time of registration. If it is not possible, choose a date that will allow for sufficient preparation time; good preparation can happen in four weeks. Actually registering for the test will make it more real and will help keep you accountable. As soon as registration is complete, open this book and *make a decision to pass the math test*!

Step 2: Get Familiar with the Test Objectives

All the necessary test skills and competencies are covered in this book. Each chapter covers a key objective. Within the framework of each chapter are more descriptive focus statements that represent the content skills needed to master each point. The questions on the math test are taken right from these descriptive statements. A detailed math lesson is provided for every focus statement, followed by sample questions with thorough explanations.

To get more familiar with the test objectives, read over the table of contents. Identifying your strengths is a perfect place to begin preparing for the test. It is helpful to place a star next to those chapter titles that you are most familiar with. After that, choose the one chapter that seems the easiest and put a double star in front of that chapter title. Next, place a check in front of any chapters that may seem troublesome.

Step 3: Know the Most Up-to-Date Format of the Test

The state certification tests are relatively new, so they are still working out the kinks. Therefore, changes in test design, requirements for passing, and format are to be expected. To get the most current information on the cost of the test, the time you are given to take it, and how many questions there are, visit your state's Educator Certification Tests website.

Step 4: Know How to Monitor and Repair Comprehension

To monitor understanding, pay close attention to the inner conversation that goes on while problem-solving. Listen to that inner voice and recognize when it is silent. When the workings of the mind are silent, there is a breakdown of comprehension. Good test-takers interact with the test. Just being aware of that will help.

To repair understanding, use fix-up tools whenever concentration or meaning breaks down. The initial and most important fix-up tool is to recognize when and how understanding wavers.

Below are some commonly used fix-up tools to repair understanding.

- Reread, reread, reread.
- Slow down.
- Always ask if what you are doing so far makes sense.
- Identify what is known and what part is confusing and when that confusion occurred.
- Do a quick summary before proceeding.
- Connect the learning to something personal. How might the problem be solved in everyday life?
- Connect the learning to something you learned at another time.
- Try to get a picture in your mind.
- Ask questions.

Step 5: Start with Your "Double Star"

In best practice, it is most effective to build on what is already known. Preparing for a test is no different. Consider your strengths and begin there. Open this book to the chapter that was identified as your "double star." This establishes a positive mindset from the beginning. Read each lesson in this chapter carefully, actively making notes in the margins. Active reading means writing down any clues, pictures, mnemonics, ideas, definitions, or connections that might help make the information come alive.

Before attempting any of the problems, reread your written notes from each lesson. This will yield a stronger memory bond so that retrieval is easier. Next, try the sample problems. Check each answer and read each explanation carefully. It is perfectly fine if you answered the problem in a different way. What is important is that the answer is right. If it is wrong, do not move on until it is clear *how* to get the right answer, and better yet, *why* the answer is what it is. If need be, go back to the lesson. Continue through the chapter using this format.

Step 6: Continue to Connect What Is Known to What Is New

The most effective study plan bridges the gap between what is known and what is new. Go back to the table of contents and choose a "one-star" chapter to continue. This will activate further background knowledge by linking your strengths to the skills necessary to pass the test. Repeat the detailed method utilized in the above "double star" chapter. It is best practice for highly effective learning to emerge.

Step 7: Take Practice Test 1

With two chapters completed, it is the perfect time to take Practice Test 1 (appendix A). With identified and strengthened assets in place, it is now critical to recognize needs. Take the practice test from beginning to end as if it were for real. Using the table provided at the beginning of each full practice test, pinpoint where the biggest need is and go to that chapter.

Step 8: Go to the Chapter That Proved to Be Most Difficult

Work slowly through the chapter that was most difficult by actively reading through each content lesson. Remember, active reading means writing down any clues, pictures, mnemonics, ideas, definitions, or connections that might help make the information come alive.

Study the examples provided. Stop before attempting the sample questions and reread the active reading notes. Rereading is a key component to comprehension. If there is any uncertainty after reading the notes written in the margins, go back and reread the lesson. This time stop after every paragraph and quickly summarize the most important idea from each paragraph. Either highlight the most important idea or write a one-sentence summary in the margin. Go back and reread the most important ideas from rereading. This time, try each sample question.

Check each answer and read each explanation carefully. Again, it is perfectly fine if the problem was solved in a different way. What is important is that the answer is right. If it is wrong, do not move on until it is clear *how* to get the right answer, and better yet, *why* the answer is what it is. If need be, go back to the lesson. Continue slowly and deliberately through the chapter using this format.

Step 9: Move through the Chapters Alternating between Easy and Hard Chapters

Using the results from Practice Test 1, go through the remaining chapters, alternating between easy and hard chapters. Make sure to use the study strategies listed above. It is critical to actively pursue clarity through the content in order to be successful. Learning is not a spectator sport. It takes work and a lot of tenacity to push through to real understanding. But it can be done, no matter how hard the material. It just takes careful planning and dedication.

Step 10: Take Practice Test 2

After finishing all the chapters in this book, take Practice Test 2 (Appendix B). Score it and determine strengths and weaknesses.

Step 11: Reflect on Results from Practice Test 2 and Make a Plan of Action

What is mastered and no longer needs a revisit? Make note of it. Writing down what is already mastered will provide proof that those skills are strong enough to pass the test. Revisit the chapters that are still problematic. Reread the areas that are most difficult and try the sample questions again. It is helpful to form a study group of peers who may also be struggling with the same material. It is beneficial to reach out to a professor or your college's academic resource center that can provide the extra boost needed to master these skills.

Step 12: Take Practice Tests 1 and 2 Again and Celebrate Progress Made

If necessary, revisit specific skills that may need further attention.

Step 13: Take Practice Test 3

Use the results of Practice Test 3 (Appendix C) to celebrate successes and determine if there are any last-minute skills that need attention. Revisit necessary areas for improvement. Recognize the effort put into this study plan and be proud. If time allows, continue to practice until the day of the test.

Step 14: You are now ready to pass the math test! Congratulations!

TAKING THE TEST: PROVEN STRATEGIES FOR SUCCESSFUL TEST-TAKING

1. Arrive to the Test Site Early

Take a practice run to the test site before the day of the test and add fifteen minutes to the traveling time on the day of the test, just in case of traffic. Show up in plenty of time to get mentally prepared, physically ready, and totally pumped to pass the test.

2. Come with a Fighter's Attitude

Sure, this test is difficult; most certification exams are. But you can take this test down and come out of it victorious. Do not be intimidated by the test. It is only a test and you are prepared to pass. Remember this fact: Expect some butterflies. Butterflies are a biological "fight or flight" response. Your body is getting ready to fight so the blood is rushing from the stomach to the legs in order to be better prepared for battle. A great strategy to help relieve this sensation is to quickly run in place and know those butterflies are working for you and not against you. Tough times come. Your body is equipped and prepared to tackle the problem. Meeting tough situations head-on is not easy, but it's all the more worth it when you achieve success in them.

Accept the fact that you will not know the answer to every question on the exam. But also accept the truth that you are capable of understanding enough about each question to make some sense of it.

A great strategy to use whenever a question seems too difficult is to follow the initial "I don't know this" comment with a "What would I do if I did know this?" response. There is something about the problem that is familiar. Find it. Perhaps there is a prefix or a root word that makes sense, a memory from an old math lesson or an experience from everyday life that will help create some clarity. Replacing the "I don't know" with "What would I do if I did know?" affords a perspective of possibility instead of defeat.

Almost every standardized test question has a bogus answer, a gotcha answer, and a just-right answer. This is valuable information to use to your advantage while taking the test. The bogus answer doesn't make sense and often doesn't even answer the question. Eliminate it right away. The "gotcha" is the answer that provides the most common error, but often still doesn't make sense to the problem. Beware of it. Asking if your answer makes sense will often eliminate the gotcha answer choice. Always look for the "just-right" answer. It will make sense, will answer the question, and will be the best-choice answer.

3. Know How Much Time You Have to Take the Exam; Plan and Pace Yourself Accordingly

Before test day, know how much time you will have for the exam. Set up a schedule for progressing through the exam. Include brain breaks along the way. (See more about the value of preplanned brain breaks below.)

During the test, work at a regular pace. Begin the exam at question 1, but feel free to skip more difficult questions in order to complete the easier ones. Just tag the unanswered questions so you can remember to get back to them. After finishing the exam, use *all* the remaining time to recheck and proofread your exam. As much as you might like to leave, force yourself to stay seated and use all the time provided.

4. Answer the Question That Is Asked

Read everything carefully and thoughtfully. It never helps to skip directions. Take your time while reading each question. Use good reading strategies throughout the test: relating things to real life, making connections to skills previously learned, re-reading, asking questions, and drawing pictures.

5. Estimate a Reasonable Answer

Estimate a reasonable answer first and then check to see which of the answer choices are close to that estimate. After working out the problem, revisit the estimate and check that the answer makes sense.

6. Use Common Sense When Eliminating Answer Choices

There is only one best answer. Analyze each answer choice as it pertains to the question. It is standard to expect that one or possibly two answers do not make sense and can be eliminated. Slow down and ask, "Does this answer make sense?" This question alone can usually narrow it down to two choices.

Often test-makers will include one "gotcha" answer choice that is the result of a common error. Playing gotcha is not nice, but it is a reality. Checking for reasonableness will lead you back into the problem to discover the error.

7. Answer Every Question

Only the number of correct answers counts. Consequently, there is no reason to leave any answers blank. Skip questions that may take too much of your time at first, but make sure to go back to them. Most of the questions will be manageable. There will be one or two that just seem impossible. You will not be penalized for incorrect answers, so it is worth guessing. Take apart what you do know about the problem. Make sense of something in the question. Narrow down the choices to two and guess.

8. Make a Decision to Stay Calm and Positive

Keep your breathing controlled and regular. This will help keep you calm. Push away any negative thoughts. Keep saying positive things to keep you mentally strong. Some good positive messages are:

- I am doing pretty well so far.
- I can do this!
- This test isn't going to get the best of me. I am a fighter!
- Stay in the game and make this work for me.

9. Use the Entire Time Provided

Use all the time provided to finish and check answers. Do not get worried if others around you leave early. That is their decision. It is helpful to identify those problems that were the most difficult. When all other questions are completed, go back to the most difficult ones and redo them, perhaps using another method to check for accuracy.

10. Maintain Mental Stamina throughout the Test

It is nearly impossible to remain focused on a single task for seventy-five minutes. Our brains just aren't wired that way. Accept that fact, and then find some way to accommodate for it. Below are three simple but very effective strategies to help maintain mental stamina:

- Include preplanned brain breaks: This will give the brain permission to recharge. Simply decide that after a given number of questions you will allow the brain the break it needs to refocus. A good analogy is to think of a boxer in a ring. After each determined round, a bell goes off and the boxer goes into his/her corner to regroup and recharge. Taking a brain break gives you that moment.
- Take your eyes off the computer, stretch in your seat, take a deep breath, look out the window, blink a few times, and think about how well you are doing so far. Then, dive back into the test. This method puts you "in the driver's seat" when it comes to staying focused. These self-determined brain breaks empower you to stay in charge. Waiting for a difficult question or a frantic moment will only complicate things and can begin a spiral downward. Be in charge of your emotions throughout the test. Before the test day, decide on a few brain breaks and build them into the test schedule. Each break will only take about thirty seconds, and to be honest, no one around you will even know. But your brain will be very grateful for it and that in turn will positively affect your score.
- Cross the midline of the body to stretch the brain: It is beneficial for thinking to do something physically before and during the test to cross the midline of the body. Before the test, touch your toes with the opposite hand. Try both sides. Now stand and stretch your opposite arm diagonally up and out. Repeat both exercises several times, while breathing deeply. Below are some quiet ways to cross the midline of the body during the test:
  ◦ Quietly stretch the left arm across the right shoulder and vice versa.
  ◦ Softly pinch your right hand to your left ear lobe, while touching your left hand to your chin. Alternate hands.
  ◦ Reach your right arm over your left shoulder, stretching down your back, while reaching your left arm up the small of your back to stretch up to reach the right arm.

Dear Future Teachers:

Best wishes on this final push toward becoming a teacher. Teaching is one of those careers that can impact our world in a big way. Purchasing this book shows your determination to do all that is possible to be part of that impact. Congratulations!

Chapter Two

Understanding Numbers and the Number System

NUMBER SYSTEMS: THE HIERARCHY OF NUMBERS

All of the numbers in this book belong to a set called *real numbers*. Real numbers are numbers that can be represented as finite or infinite decimals and can be placed on the number line. They are either rational or irrational. A finite decimal is a decimal that terminates or ends. Examples of finite decimals are 0.625 and 0.9387124. An infinite decimal is a decimal that goes on forever and the numbers in the decimal may or may not form a repeatable pattern. An example of an infinite decimal that does not repeat is 0.2317493628.... The three dots (ellipsis) indicate the decimal goes on. An example of an infinite decimal that repeats is the 0.33333.... A repeated decimal like this is usually rounded, truncated (cut off), or written with a bar over the repeating numbers to show that it repeats. $0.\overline{3}$

Irrational numbers are real numbers that are nonrepeating, nonterminating decimals and cannot be written in the form $\frac{a}{b}$ where both a and b are integers and $b \neq 0$.

Rational numbers are real numbers that can be written as the ratio (fraction) of two integers ($\frac{a}{b}$), where the denominator of the ratio cannot be 0. As decimals, they are either terminating or repeating.

Natural or counting numbers are the set of whole numbers starting with 1 and increasing infinitely.

Whole numbers are the set of integers beginning at 0 and increasing infinitely. All natural numbers or counting numbers are whole numbers.

Integers are rational numbers that include the positive whole numbers, their opposites, and 0. They do not include any measures between whole numbers. All whole numbers are integers.

A *prime number* is a natural number greater than 1 that has only itself and 1 as factors. 2 is the only even prime number. All other even numbers are not prime because they also contain a factor of 2.

A *composite number* is a natural number greater than 1 that is divisible by more than itself and 1. The number 1 is neither prime nor composite.

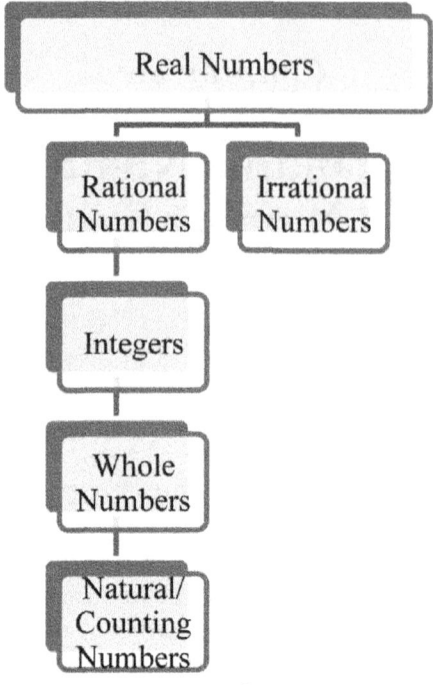

Figure 2.1. Hierarchy of Numbers

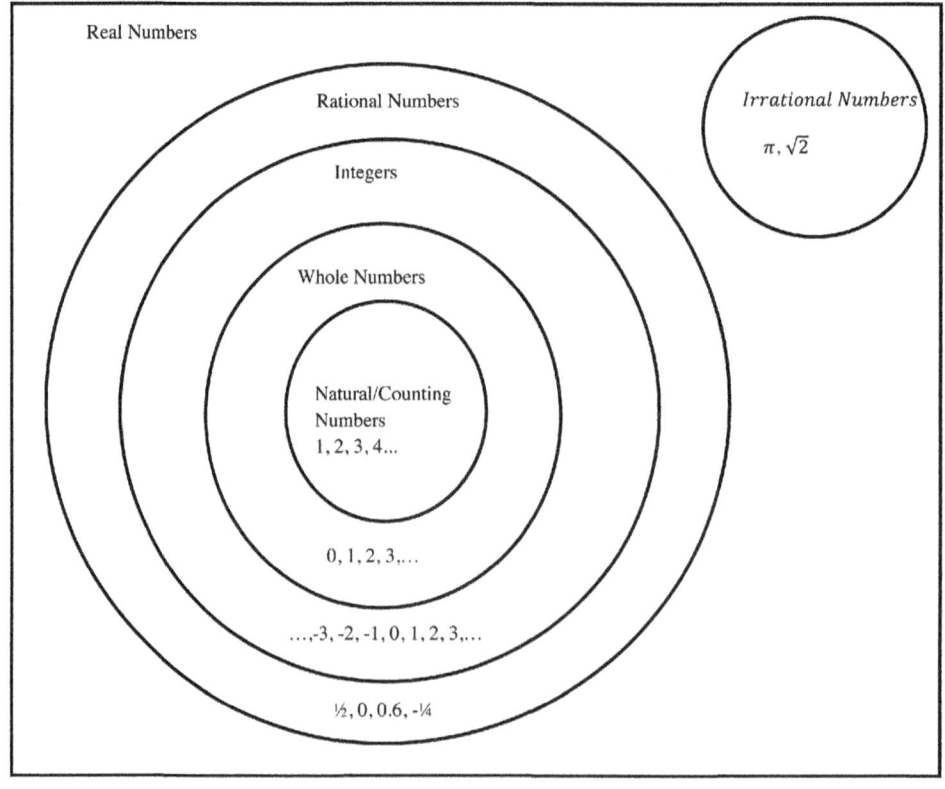

Figure 2.2. Concentric Ovals of Rational Numbers

Example 1

List elements of the set of integers between −7 and 4.
 Answer: {−6, −5, −4, −3, −2, −1, 0, 1, 2, 3}
 Integers are all whole numbers with their opposites and 0. −7 and 4 are not included in the solution. Whenever the word *between* is used, it means not to include the numbers that the answer is between. When the word *inclusive* is used, it means to include the number(s) mentioned.

Example 2

The product of two consecutive, even integers is 2,208. What are the integers?
 Answer: 46, 48
 Consecutive means "in order" and even numbers are numbers that end in 0, 2, 4, 6, or 8. An example of consecutive even integers is 12 and 14. The word *product* means the answer of a multiplication problem. Next, check out the digits in the ones place—notice that both 2 × 4 and 6 × 8 result in an 8 in the ones place, so now use trial and error to fill in likely tens place values. 22 × 24 = 528, which is way too small, so try 26 × 28 = 728. Next, try 32 × 34 = 1,088 and 36 × 38 = 1,368, which is getting closer. The next two consecutive, even integers to try are 42 × 44 = 1,848 and finally 46 × 48=2,208.

Practice Test Questions on Number Systems

1. List elements in the set of natural numbers less than and including 5.
   (a) {0, 1, 2, 3, 4, 5}
   (b) {1, 2, 3, 4, 5}
   (c) {0, 1, 2, 3, 4}
   (d) {1, 2, 3, 4}
2. List elements in the set of integers less than 2.
   (a) {. . . −2, −1, 0, 1}
   (b) {−2, −1, 0, 1, 2}
   (c) {−2, −1, 0, 1, 2}
   (d) {. . . −2, −1, 1}
3. A number of the form $2^n - 1$ is called a Mersenne Prime, where n is any natural number. The result is often a prime number, but it is not true for all numbers. Which number below when placed in the expression will not result in a prime number?
   (a) 2
   (b) 3
   (c) 6
   (d) 4
4. A teacher states that the product of four whole numbers cannot be negative. Which is proof that the teacher is right?
   (a) (−5)(−6)(−2)(−1)
   (b) 3 × 5 × 7 × 2
   (c) (−4) + (−5) + (−2) + (−6)
   (d) 3 + 5 + 7 + 2

Detailed answer key for the above problems:

1. Answer: (b) {1, 2, 3, 4, 5}
   Natural numbers do not include 0. 5 is also included as part of the solution set.
2. Answer: (a) {. . . −2, −1, 0, 1}
   This means all integers less than 2. 2 is not included as part of the solution set because it states "less than 2." 0 is included because 0 is an integer.
3. Answer: (d) 4
   First, substitute each of the values for n. (a) $2^2 - 1 = 3$. 3 is prime. (b) $2^3 - 1 = 7$. 7 is prime. (c) $2^6 - 1 = 31$, and 31 is prime. The answer should then be (d) $2^4 - 1 = 15$, and 15 is not prime. 15 is composite because it has more than itself and 1 as factors. The factors of 15 are 1, 3, 5, and 15.
4. Answer: (b) $3 \times 5 \times 7 \times 2$
   $3 \times 5 \times 7 \times 2$ is correct because it is the only choice that includes both multiplication and whole numbers. Choice (a) does result in a positive product, but the factors are integers and not whole numbers.

DECIMALS

The decimal system is a base-ten place value system. The prefix *deci-* means $\frac{1}{10}$, so every time a place value moves from left to right in the decimal system, its value is divided by 10 and conversely, every time a place value moves from right to left in the decimal system, that value is ten times greater. For example in the number 2,547, the value of 2 is two thousand, the value of 5 is five hundreds, the value of 4 is four tens, and the value of 7 is seven ones.

Rounding Decimals

Rounding decimals is very similar to rounding whole numbers, except drop the zeroes at the end of a decimal number because it does not change the value of a number.

Table 2.1. Decimal Place Value Chart

Decimal Place Value										
thousands	hundreds	tens	ones	.	tenths	hundredths	thousandths	ten-thousandths	Hundred-thousandths
3,	4	2	1	.	0	0	2	1	6

Three thousand four hundred twenty-one and two hundred sixteen hundred-thousandths

Example 1

Round 3.2562459 to the nearest hundredths.

1. Underline the place value you are rounding to: 3.2<u>5</u>62459.
2. Put a circle around the number to the right of that digit. This number is your guide number. The rounding poem helps to remember the process: "4 or below let it go; if it is 5 or above, give it a shove." Notice that the guide number 6 is 5 or above, so give the digit in the hundredths place a shove up to 6. 3.2<u>5</u>⑥2459
3. Now all the remaining decimal numbers to the right of the hundredths place get dropped and every number to the left of the hundredths place stays the same. 3.26.
4. Dropping zeroes when rounding money to the nearest cent is a little different. Keep two decimal places to show dimes and pennies, even if they are zeroes. Example: using money and rounding to the nearest cent. $3.202563 rounded to the nearest cent is $3.20.

SCIENTIFIC NOTATION

Scientific notation is a way to write really, really big numbers and really, really small numbers without using all the zeroes they contain. In fact if 35,000,000,000 and 1,400,000,000,000 are multiplied using a calculator, the answer usually shows as: 4.9E22 which really means 4.9×10^{22}. The calculator could not display this really big number, so it converted it into scientific notation for you—hence the name "scientific calculator."

A number that is in scientific notation is a product of a number greater than or equal to 1 but less than 10 and an integer power of 10. To write a number in scientific notation, move the decimal point to the right of the first digit in the numbers. Count how many places the decimal point has been moved. This number of times the decimal point was moved will be the exponent. The exponent will end up being positive if the decimal point was moved left and negative if the decimal point was moved right.

17,000,000 written in scientific notation is 1.7×10^7

.000000056 written in scientific notation is 5.6×10^{-8}

Example 1

In the number 982,436,200.521, which of the following is the sum of the hundredths place and the millions place?

(a) 7
(b) 4
(c) 3
(d) 9

Answer: (b) 4

The digit in the hundredths place is 2 and the digit in the millions place is 2. The sum of 2 and 2 is 4.

Example 2

After dividing 4.62×10^6 by four, what digit would be in the hundred thousands place?

(a) 0
(b) 4
(c) 1
(d) 6

Answer: (c) 1
$4.62 \times 10^6 = 4,620,000$. Divide that number by 4. The answer is 1,155,000. 1 is the digit in the hundred thousands place.

Practice Test Questions on Decimals

1. Which of the following is the product of $(1.2 \times 10^4) \times (3.1 \times 10^8)$?
   (a) 3.72×10^{12}
   (b) 3.4×10^{12}
   (c) 372×10^{12}
   (d) 6.51×10^{14}
2. Subtract 0.00009 from 0.000113. What would that difference be in scientific notation?
   (a) 2.3×10^{-5}
   (b) -2.3×10^{-5}
   (c) 1.017×10^{-8}
   (d) 0.000023
3. What is the value of 4 in the number 6.42×10^8?
   (a) 4 hundred thousands
   (b) 4 million
   (c) 4 ten millions
   (d) 4 tens
4. Which of the following is equal to a quarter of a billion?
   (a) 250,000
   (b) 2,500,000
   (c) 0.2500000
   (d) 250,000,000

Detailed answer key for the above problems:

1. Answer: (a) 3.72×10^{12}
   There are two easy ways to find this answer. Find each product ($1.2 \times 10^4 = 12,000$ and $3.1 \times 10^8 = 310,000,000$). Multiply 12,000 and 310,000,000 to get 3.72×10^{12}. Another option is to multiply $1.2 \times 3.1 = 3.72$ and $10^4 \times 10^8 = 10^{4+8} = 10^{12}$ and simply leave them as factors, leaving the answer 3.72×10^{12}. See the section on exponents in this chapter for more information on the rules of exponents. Answer (c) is incorrect because scientific notation can only have one leading digit.

2. Answer: (a) 2.3×10^{-5}
   In subtraction, the "from" number always goes first. Subtract $.000113 - .00009 = .000023$ which is 2.3×10^{-5}. The calculator might show 2.3E–5, which means 2.3×10^{-5}. Be careful to subtract 0.00009 from 0.000113. Answer choice (b) is the result of subtracting in the wrong order. This choice is not in scientific notation anyway because the first factor is not between 1 and less than 10.
3. Answer: (c) 4 ten millions
   Multiply $6.42 \times 10^8 = 642,000,000$. Follow place values from right to left and identify the 4 in the ten millions place.
4. Answer: (d) 250,000,000
   This is really the only answer that makes sense because when dividing 1,000,000,000 by 4, the result would still have to be a really big number. None of the other options are close. It is also a good idea to use the answer choices provided and multiply each option by 4 to get to the answer. $250,000,000 \times 4 = 1,000,000,000$.

PERCENTS

The word *percent* means parts per hundred. A percent is a way to show a relationship between part of a whole and the whole. The whole in a percent is always out of 100. The symbol % means "per hundred."

Connecting Percents to Fractions to Decimals

Percents, fractions, and decimals all describe parts of a whole. There are three ways of representing the same amount. For example, if your name is "Margaret" and your friends call you "Margie" and your family calls you "Marge," you are still the same person—but you have three names. The fraction $\frac{1}{2}$, the decimal 0.5, and the percent 50% all mean the same amount, but are three different ways of expressing it. Expressing different ways to show equivalent numbers is easy:

1. Converting decimals to percents—simply multiply the decimal by 100.
   Example: Convert 0.245 to a percentage. Multiply 0.245×100 because *percent* means "out of 100." 0.245 as a percent is 24.5%
2. Converting percents to decimals—simply divide the percent by 100.
   Example: Convert 12% to a decimal. 12% divided by 100 is 0.12.
   To check this, do the opposite operation to make sure to get the original number by multiplying 0.12 by 100, which is 12%.
3. Converting fractions to decimals—since every fraction bar is really read "divided by," then simply divide the numerator by the denominator.
   Example: Convert $\frac{3}{5}$ to a decimal. $\frac{3}{5}$ can be read 3 divided by 5. To do this, first tip the numerator 3 off the top of the fraction and let it fall into a division house as the dividend and leave the denominator outside as the divisor.

$$3 \div 5 = 0.6.$$

$$\frac{3}{5} = 0.6$$

4. Converting fractions to percents—do step 3 above; then once the result is a decimal, complete step 1 above.
   Example: Convert $\frac{4}{5}$ to a percent. $4 \div 5 = 0.8$. Multiply $0.8 \times 100 = 80\%$.
5. Converting decimals to fractions simplified—it is most important to first understand decimal place value and to be able to read decimals by their value. The decimal 3.25 is read three and twenty-five hundredths and not read as "three point twenty-five." The decimal point is always read as "and."
   Example: Convert 0.125 as a fraction simplified. 0.125 read out loud is "one hundred twenty-five thousandths." Now picture what that looks like as a fraction. The last digit read will identify the place value of the denominator: $\frac{125}{1000}$. Read both decimal and fraction again and make sure they are saying the same word name. The last step is to simplify the fraction by dividing both numerator and denominator by common factors until the only factor they share is 1. Both 125 and 1,000 share a factor of 25. Divide both 125 and 1,000 by 25 and get the fraction $\frac{5}{40}$. This can be simplified again by dividing 5 into the numerator and denominator. $0.125 = \frac{1}{8}$. Dividing by a greatest common factor (GCF) will simplify fractions in one step. The GCF of 125 and 1,000 is 125, so dividing both numerator and denominator by 125 will get $\frac{1}{8}$ in one step. But it doesn't matter if fractions are simplified in one step or a few steps; it just matters that the fraction is in its simplest form.

Example 1

Suppose you convert 0.28 to a fraction simplified; what would be the denominator simplified?

Answer: 25.

Read 0.28 as twenty-eight hundredths. Write that as a fraction: $\frac{28}{1000}$. Reread both the decimal and fraction to make sure they are saying the same word name. If so, continue with your procedure. Finally, check that the fraction is in its simplest form. Divide both numerator and denominator by 4, which simplified is $\frac{7}{25}$. The denominator is 25.

Example 2

Compute 1.145 − 0.995. Convert the difference as a fraction simplified. What would be the resulting denominator?

Answer: 20

$$1.145 - 0.995 = 0.15$$

Convert 0.15 as a fraction by reading it as fifteen hundredths. Write it as a fraction: $\frac{15}{100}$. Read the decimal and fraction aloud to make sure they say the same word name. Simplify the fraction by dividing both numerator and denominator by their GCF, which is 5. The simplified fraction is $\frac{3}{20}$. The denominator is 20.

Finding the Percent of a Number

What is 52% of 82?

1. First estimate 50% (or $\frac{1}{2}$) of 82, which is 41. Since 52% is slightly more than $\frac{1}{2}$, the answer should be slightly more than 41.
2. Whenever you are taking a value *of* something, it often means multiplication. Convert 52% to a decimal and multiply the two factors: $0.52 \times 82 = 42.64$.
3. Revisit the original estimate. Does the answer make sense? Yes, it is a little bigger than 41.

Example 1

At Cabrini College, 60% of the students live on campus. If there are 1,400 students going to Cabrini College, how many live on campus?
   Answer: 840 students live on campus

1. The first thing to do is to estimate. 60% is greater than 50%, so the answer will be greater than 700, and since 60% is closer to 50% than it is closer to 100%, the answer should be closer to 700 than 1,400. This will start you off with a good estimate.
2. Recognize the word *of* in the problem. If you are taking 60% of anything, you will need to multiply. Now convert 60% to a decimal by using step 2 above and multiply it (.60) by 1,400. $0.6 \times 1400 = 840$.
3. Ask yourself, does this answer make sense? Is it relatively close to my estimate? Yes it is.
4. There are 840 students who live on the Cabrini College campus.

Example 2

At Springfield High School, 40% of the students take geometry. There are 480 students who take geometry. How many students are there in all at Springfield High School?
   Answer: 1,200

1. Realize that this problem is asking for the total amount of students and gives the percentage of students who take geometry. So, if the first example requires multiplication, this example will require the inverse of multiplication, which is division.
2. Estimate by picturing if 40% of something is 480, what might that something be? It would be bigger than doubling 480 because that would make 80% of the total. So if 480 only represents 40% and not 50%, the total number of students will be greater than 960.
3. Set up an equation that represents what you know and include a variable for what you need to know: 40% of x = 480.
4. Convert 40% to a decimal and create an equation.
5. $0.4x = 480$

6. Divide 480 by 0.4.
7. There are 1,200 students in all.

OR

1. Convert the 40% to a fraction simplified and create an equation.
2. $\frac{2}{5}x=960$.
3. Multiply 960 × 5 and then divide that product by 2.
4. There are 1,200 students.
5. Revisit the original estimate and ask if the answer makes sense. Yes, it is close to the original estimate.

Finding a Percent Increase or Decrease

To find a change in percents (either an increase or decrease), use the following formula:

$$\frac{\text{New Value} - \text{Old Value}}{\text{Old Value}} \times 100$$

If the answer is a negative number, it represents a percent decrease; if it is positive, the answer represents a percent increase.

Practice Test Questions on Converting Numbers and Working with Percents

1. 1 divided by 4 is equivalent to which of the following?
   i. 0.25
   ii. $\frac{1}{8} \div 2$
   iii. $\frac{1}{8} \div \frac{1}{2}$
   (a) (i) only
   (b) (i) and (ii)
   (c) (i) and (iii)
   (d) (ii) only
   (a) 1,600,000
   (b) 160,000
   (c) 220,000
   (d) 8

Table 2.2. TV Show Ratings by Genre in 2012

Genre of TV Show	American Viewers in 2012
Reality Show	30 percent
Talent Competition	40 percent
Comedy Sitcom	8 percent
Crime/Police Drama	22 percent

2. Use the table on the preceding page to answer the following question:
   Last year 2,000,000 American viewers were asked which genre of show they preferred to watch on TV. How many viewers said they preferred comedy sitcoms?
3. Use the table below to answer the following question:
   What percent of the population of bilingual citizens of Texas live in Austin?

Cities in Texas	Percent of Bilingual Citizens
Austin	19.3
Dallas	17.5
El Paso	53.8
Galveston	24.6

   (a) 19.3%
   (b) 1,930%
   (c) 16.75%
   (d) 20%
4. Every year after Christmas, Bodine TV & Appliances has a storewide sale. Last year a standard thirty-one-inch TV that normally sells for $600 was on sale for $360. What is the percent of decrease in the sale for the thirty-one-inch TV?
   (a) 50%
   (b) 60%
   (c) $50
   (d) 40%

Detailed answer key for the above problems:

1. Answer: (c) both (i) and (iii)
   $\frac{1}{4} = 0.25$, so (i) has to be an option. This eliminates choice (d).
   $\frac{1}{8} \div 2 = 0.0625$, which does not equal $\frac{1}{4}$ so (ii) is not an option. This eliminates choice (b).
   $\frac{1}{8} \div \frac{1}{2} = \frac{1}{4}$, so both (i) and (iii) are equal to $\frac{1}{4}$. (c) is the answer.
2. Answer: (b) 160,000
   Use the table to identify that 8% of the 2,000,000 viewers prefer comedy sitcoms. Estimate 8% of 2 million. Since 8% is a little less than 10%, the answer should be a less than 200,000.
   Convert 8% as a decimal by dividing it by 100 to get 0.08. Finding 0.08 of 2,000,000 indicates multiplication. $0.08 \times 2,000,000 = 160,000$.
   Revisit the estimate to see if the answer makes sense. Yes, choice (b) is the only answer that is close to but less than 200,000.
3. Answer: (a) 19.3%
   Read the table to identify that the data provided is already in percent form. Therefore, the answer 19.3 for the percentage of citizens in Austin, Texas, who are bilingual is 19.3%.

20 • Chapter Two

4. Answer: (d) 40%

First, find the rate of discount for last year's sale.

$$\frac{\text{New Value} - \text{Old Value}}{\text{Old Value}} \times 100$$

Substitute what is known for the variables:

$$\frac{360 - 600}{600} \times 100$$

Since the result of the division problem is negative, the quotient represents a percent of decrease. Disregard the negative and multiply 0.4 by 100 to convert the answer to a percent of decrease.

SIMPLE INTEREST AND COMPOUND INTEREST

Simple Interest

Interest is the amount paid for the use of money. For example, when money is deposited in a savings account, interest is earned on it. When money is borrowed, interest is what is paid on what was originally borrowed. The amount of simple interest paid on a savings account or charged on a loan depend on the rate of interest, the amount saved or borrowed (the principal), and the length of time that money is in the account or the loan is outstanding.

The following formula is used when calculating simple interest:

$$I = P \times r \times t$$

The I or A stands for the *interest*, the amount paid for the use of the money.
The P stands for the *principal* that was originally borrowed or deposited.
The r stands for the *rate* or percent of interest paid or earned.
The t stands for the number of years or fractional part of a year that the money is borrowed or deposited.

Example 1

Christopher is taking a graduate course in the fall. The tuition costs $2,800, which he borrows from the college at a simple interest rate of 2.5%. If Christopher pays the loan back in six months, how much will he owe the college in all, including the interest?

Answer: $2,835

Use the above formula: $I = P \times r \times t$

Convert the rate to a decimal by dividing by 100 or simply moving the decimal point two places to the left. Convert the six months to represent a decimal or fractional part of a year. Plug in what is known to calculate the interest: $I = 2,800 \times .025$

× .5=35. The interest Christopher has to pay back in six months for the loan is $35. The total amount he will pay for the course is $2,835.

Example 2

Drew borrows $5,000 for a used truck. He plans to repay the loan in 2 years at simple interest. If he repays a total of $5,400, what is the rate of interest?
Answer: 4%
Use the simple interest formula: $I = P \times r \times t$
First calculate the interest by subtracting the amount borrowed from the future value of the loan: $5,400 - $5,000 = $400 interest paid. Then plug in the values known to calculate the unknown rate.

$$400 = 5,000 \times r \times 2$$

Solve for the rate.

$$400 = 10,000r$$

Divide both sides of the equation by 10,000 to calculate the rate.

$$400 \div 10,000 = .04$$

Convert 0.04 to a percent by multiplying it by 100 or simply moving the decimal point two places to the right. Drew is paying 4% interest on his loan for the truck.

Compound Interest

Compound interest is exactly what it sounds like: interest that is calculated on the original principal, "compounded" with the accumulated unpaid interest. It is the cost of the interest on the principal and the interest accumulated to date. Compound interest involves the same factors as simple interest, but it is calculated with the principal and the accumulated interest taken together at the end of each pay period. Most saving accounts pay compound interest; that is, interest paid on the principal as well as on the accumulated interest. Interest can be "compounded" for any amount of time (semi-annually, quarterly, etc.).

The following formula is used when calculating compound interest:

$$A = P(1+r)^t$$

The A stands for the value of the interest at time t, or the future value.
The P stands for the *principal* that was originally borrowed or deposited.
The r stands for the *rate* or percent of interest paid or earned.
The t stands for the number of years that the money is borrowed or deposited.

Example 1

Andy deposits $12,000 into a CD that earns 4% interest compounded annually. At the end of five years, what is the future value of Andy's account? Round to the nearest dollar.

Answer: Use the formula above to substitute each of the given values.

$$A=P(1+r)^t$$
$$A=12,000(1+.04)^5$$

The A or future value is unknown.
The P or *principal* is $12,000.
The r stands for the *rate*, which is 4% converted to a decimal: 0.04.
The t stands for the number of years. In this case, Andy is depositing the money for five years.
Use order of operations. Simplify inside the parentheses.

$$1 + .04 = 1.04$$

Calculate $1.04^5 = 1.216652902$.
Multiply that *by* 12,000 *to get* $14,599.83483.
Round this answer to the nearest dollar to get $14,600; the value of the CD after five years will be $14,600.

Example 2

Which of the following expressions describes a $2,000 investment at 4.2% interest compounded annually for ten years?

(a) $2,000 \times (.042)^{10}$
(b) $2,000 \times (1.042)^{10}$
(c) $2,000 \times .042 \times 10$
(d) $2,000 \times 1.042 \times 10$

Answer: (b) $2,000 \times (1.042)^{10}$
Use the formula to substitute the given values in order to discover the appropriate expression:

$$A=P(1+r)^t$$
$$A=2,000(1+.042)^{10}$$

The A or future value is unknown.
The P or *principal is* $2,000.
The r stands for the *rate*, 4.2%, which converted to a decimal is 0.042.
The t stands for the number of years. In this case, Andy is depositing the money for ten years.
Thus, the only choice that makes sense is (b) $2,000 \times (1.042)^{10}$

Practice Test Questions on Simple and Compound Interest

1. Gloria deposits $750 into a savings account that earns 2.5% simple annual interest. What is the balance in the account after eighteen months? Round to the nearest cent.
   (a) $778.13
   (b) $28.13
   (c) $721.87
   (d) $750 × .025 × 1.5
2. Jason takes out a loan for $8,000. The bank charges 6.5% annual simple interest. What is the interest Jason will pay on the loan if he pays it off in 6 years?
   (a) $11,120
   (b) $200
   (c) $8,000 × .065 × 6
   (d) $3,120
3. Which of the following expressions would be best to calculate the future value of $5,000 invested at 1.25% interest, compounded annually for five years?
   (a) 5,000 × .0125 × 5
   (b) 5,000(.0125 × 5)
   (c) $5,000(1.0125)^5$
   (d) $5320.41
4. Ayden borrowed $7,000 at an interest rate of 7% per year. If he paid $1,960 in simple interest, for how many years did he take out the loan?
   (a) 1
   (b) 2
   (c) 3
   (d) 4

Detailed answer key for the above problems:

1. Answer: (a) $778.13
   Since it is a simple interest problem, use the above simple interest formula: $I = Prt$. Convert the rate to a decimal by dividing by 100 or simply moving the decimal point two places to the left: 2.5% = 0.025. Convert the eighteen months to represent a decimal or fractional part of a year: 18 months = 1.5 years. Plug in what is known to calculate the interest: I = 750 × 0.025 × 1.5 = 28.125. Round this to the nearest penny, which is $28.13. Add the interest to the principal to calculate the balance of the account: 750 + 28.13 = $778.13. Answer choice (b) is wrong because it is the interest only, without the principal. It does not answer the question. Answer choice (c) would have been correct if it were money borrowed and the interest would then have to be deducted from the principal. Answer choice (d) provides the expression only, which does not answer the question.
2. Answer: (d) $3,120
   Since it is a simple interest problem, use the above simple interest formula: $I = Prt$. Convert the rate to a decimal by dividing by 100 or simply moving the decimal point two places to the left: 6.5% = 0.065. Use six years in for the time. Plug in what is known to calculate the interest:
   I = 8,000 × .065 × 6 = $3,120.

Answer choice (a) is wrong because it is the future value of the loan. It does not answer the question. Answer choice (b) makes no sense in the problem and can be eliminated. Answer choice (c) provides the expression only, which does not answer the question.

3. Answer: (c) $5,000(1.0125)^5$

   Use the formula to substitute the given values in order to discover the appropriate expression:

   $$A = P(1+r)^t$$
   $$A = 5,000(1+.0125)^5 \text{ or } A = 5,000(1.0125)^5$$

   The A or future value is unknown.
   The *principal* is $5,000.
   The r stands for the *rate*, which is 1.25% compounded annually—converted to a decimal is 0.0125.
   The t stands for the number of years. In this case, the time is five years.
   Thus, the only choice that makes sense is (c) $A = 5,000(1+.0125)^5$ or $A = 5,000(1.0125)^5$.

4. Answer: (d) 4

   Since it is a simple interest problem, use the above simple interest formula: $I = Prt$. Convert the rate to a decimal by dividing by 100 or simply moving the decimal point two places to the left: 7% = 0.07. The principal money Ayden borrowed is $7,000. Substitute that in for P, principal. This time the interest is also known: interest = $1,960. The time is unknown. Plug in the given values to determine the time.

   $$1,960 = 7,000 \times .07 \times t$$

   Solve for t.

   $$1960 = 490t$$

   Divide both sides of the equation by 490 to find the value of t.

   $$t = 4 \text{ years}$$

EXPONENTIAL NOTATION

Exponential notation is a way of writing repeated multiplication of the same number using exponents. Example $3^4 = 3 \times 3 \times 3 \times 3 = 81$

The operation used is always multiplication.

An *exponent* is a little number that sits on the upper right shoulder of a normal-sized number. The exponent identifies how many times to use the base number as a factor of itself. 2^3 is read either "two to the third power" or "two cubed." The exponent 3 means to use 2 three times as a factor, $2 \cdot 2 \cdot 2 = 2^3 = 8$.

Any number written to the zero power is equal to 1: $5^0 = 1$; $121^0 = 1$; $1,00000^0 = 1$; etc.

Same-base product rule: When multiplying exponential terms whose bases are the same, keep the base and add the exponents. Example: $5^2 \times 5^4 = 5^{2+4} = 5^6 = 5 \cdot 5 \cdot 5 \cdot 5 \cdot 5 \cdot 5 = 15{,}625$.

Same-base quotient rule: When dividing exponential terms whose bases are the same, keep the base and subtract the exponents. Example: $9^5 \div 9^3 = 9^{5-3} = 9^9 = 81$.

Negative exponents: A number raised to a negative power can be restated the number as a denominator of a fraction making its exponent positive and placing a one in the numerator. Example of a negative exponent: $2^{-2} = \frac{1}{2^2} = \frac{1}{4}$

Example 1

Simplify $\frac{(-5)^2}{(-5)^2}$
  Answer: $(-5)^{2-2=0}$; $(-5)^0 = 1$

Example 2

Simplify $6x^0 y^{-4} \cdot 2y^5$.
  Answer: Multiply the constants: $6 \cdot 2 = 12$.
  Simplify $x^0 = 1$.
  Simplify $y^{-4} \cdot y^5 = y^{-4+5} = y^1$.
  $6x^0 y^{-4} \cdot 2y^5$ simplified equals $12y$.

Practice Test Questions on Exponents

1. Write the following numbers from least to greatest: 2^3, 3^2, 1^{14}, 4^{-1}.
   (a) 2^3, 3^2, 1^{14}, 4^{-1}
   (b) 1^{14}, 2^3, 3^2, 4^{-1}
   (c) 4^{-1}, 1^{14}, 2^3, 3^2
   (d) 24^{18}
2. Simplify $(-s)^2 \cdot (-s)^3 \cdot (-s)^4$.
   (a) s^9
   (b) $(-s)^9$
   (c) s^{24}
   (d) $(-s)^{24}$
3. Simplify $7y^0 x^4 \cdot 2x^2$.
   (a) $14x^6$
   (b) $14y^0 x^6$
   (c) $14x^8$
   (d) $14y^0 x^8$
4. Simplify 2^{-5}
   (a) $-\frac{1}{32}$
   (b) $\frac{1}{32}$
   (c) $-\frac{1}{10}$
   (d) 32

Detailed answer key for the above problems:

1. Answer: (c) 4^{-1}, 1^{14}, 2^3, 3^2

$$4^{-1} = \tfrac{1}{4}$$
$$1^{14} = 1$$
$$2^3 = 2 \cdot 2 \cdot 2 = 8$$
$$3^2 = 3 \cdot 3 = 9$$

2. Answer: (b) $(-s)^9$
To simplify $(-s)^2 \cdot (-s)^3 \cdot (-s)^4$, add the exponents and keep the base. $(-s)^{2+3+4} = (-s)^9$. (a) is incorrect because the base must stay the same. (c) is incorrect because the exponents are multiplied instead of adding them and the base is wrong. (d) is incorrect because the exponents are multiplied instead of adding them.

3. Answer: (a) $14x^6$
To simplify $7y^0 x^4 \cdot 2x^2$, multiply the coefficients: $2 \cdot 7 = 14$.
Simplify y^0 as 1, which is understood to be there, but does not need to be written.
Simplify $x^4 \cdot x^2 = x^{4+2} = x^6$.
The expression $7y^0 x^4 \cdot 2x^2$ simplified is $14x^6$.

4. Answer: (b) $\tfrac{1}{32}$

$$2^{-5} = \tfrac{1}{2^5} = \tfrac{1}{32}$$

Any negative exponent must be shown as a fraction with one as the numerator and the base and positive power as the denominator.

Chapter Three

Pre-Algebra

ANALYZING AND EXTENDING A VARIETY OF PATTERNS

This type of problem involves looking at charts or pictures, figuring out what changes have occurred from one row or picture to the next, and then projecting that change forward to the next row or picture. Some typical problems involve adding the same number to the previous number, adding the two previous numbers to get the next number, or multiplying the previous row by some number to get the next number.

Approach these problems by reading carefully and making up a chart. Notice whether the numbers are increasing or decreasing. Write down the information that you know. Look at the numbers and figure out how they are changing. Be careful and do not jump to conclusions too quickly. Be sure to test the pattern moving from each number to the next.

Example 1

A piece of paper is cut into 3 rectangles. Each of those rectangles is cut into 3 pieces. How many rectangles will there be after the 5th cutting?

Answer: 243 rectangles

Start by setting up a table.

Cut	Original	After cut
1st	1	3
2nd	3	9
3rd	9	27
4th	27	81
5th	81	243

The original number of rectangles is multiplied by 3 to get the new number of rectangles. Continuing that pattern to the 5th cut will result in 243 rectangles.

Example 2

3, 10, 4, 12, 5, . . . What is the next number in the sequence?
Answer: 14

The 1st, 3rd, and 5th numbers are increasing by 1. The 2nd and 4th numbers are increasing by 2. So, the 6th number will also increase by 2 and the next number in the sequence will be 14.

Example 3

How many dots will be on the sixth block?
Answer: 20 dots

The 1st block has 4 dots and the 2nd block has 6 dots, so moving from the 1st block to the 2nd block, 2 dots were added. The 3rd dot has 9 dots, so moving from the 2nd block to the 3rd block, three dots were added. It looks like the pattern is that the number of dots being added is getting greater by 1 each time. This means that the 4th block should have 4 dots added and therefore have 13 dots: 9+4=13. But, there are only 12 dots on the 4th block. So, the 1st guess at the pattern is incorrect. Closer examination shows that when moving from the 1st block to the 2nd block, a new horizontal row was added. A new vertical column was added moving from the 2nd block to the 3rd block. Then a row was added on the 4th block. So, a column will be added to the 5th block and a row will be added to the 6th block. This results in a block with 20 dots.

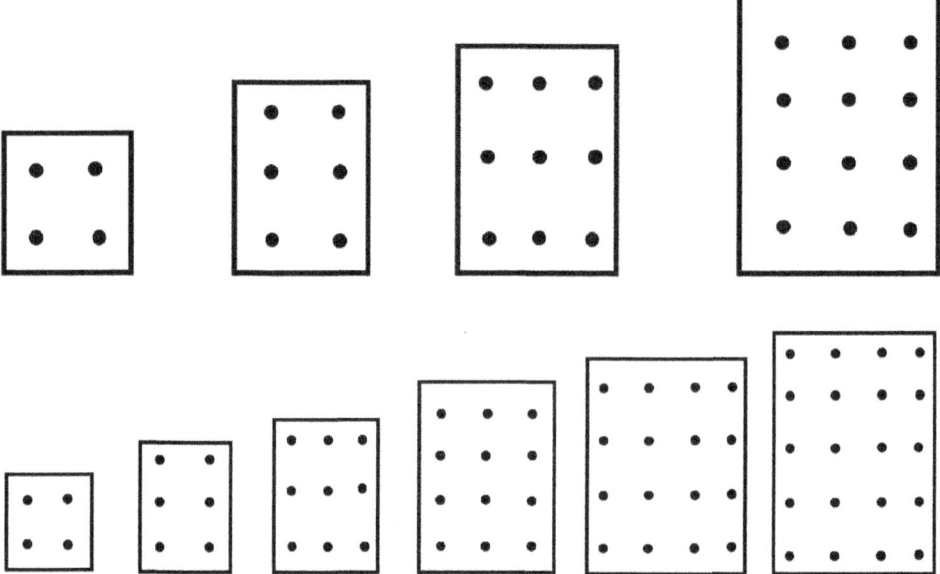

Pre-Algebra

Practice Test Questions on Analyzing and Extending a Variety of Patterns

1. Pete likes to combine his math skills with his eating habits. On the 1st day of the month, he eats 1 pea and 2 kernels of corn. On the 2nd day of the month, he eats 2 peas and 4 kernels of corn. On the 3rd day of the month he eats 3 peas and 6 kernels of corn. He continues this pattern for the month of July (31 days). How many peas and how many kernels of corn will Pete eat on July 31st?
   (a) 30 peas, 60 kernels of corn
   (b) 31 peas, 62 kernels of corn
   (c) 62 peas, 31 kernels of corn
   (d) 31 peas, 93 kernels of corn

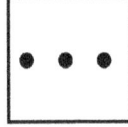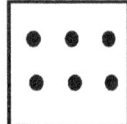

2. What kind of figure will have 20 dots?
   (a) circle
   (b) triangle
   (c) square
   (d) none of them will have 20 dots

3. Using the table below, identify what operation(s) is always done to each number in the 1st column to get each number in the 2nd column. What will the 2nd number be if the 1st number is 41?
   (a) add 2 to the previous second number; 13
   (b) add 1 to the previous first number; 42
   (c) multiply the first number by 2 and then add 7; 89
   (d) there is no pattern

x	y
-1	5
0	7
1	9
2	11

4. What is the ones digit in the 12th power of 8?
   (a) 6
   (b) 0
   (c) 4
   (d) 8

Detailed answer key for the above problems:

1. Answer: (b) 31 peas, 62 kernels of corn
   Use a chart to solve this problem. Fill in the chart with the given information.

Day	Peas	Corn
1	1	2
2	2	4
3	3	6
4	4	8

   Then look at the chart and see if there is a pattern. The problem can be solved by extending the chart and continuing the pattern (increase the number of peas by 1 and increase the number of kernels of corn by 2) until the 31st day is reached. This will take a long time! Another approach is to look for the relationship between the day number, the number of peas, and the number of corn kernels. The number of peas is the same as the day number. The number of corn kernels is 2 times the day number. So, on the 31st day, Pete will eat 31 peas and 62 kernels of corn. Using the relationship to find the answer will be faster than extending the chart.

Day	Peas	Peas/Week	Corn	Corn/Week
1	1	1	2	2
2	2	3	4	6
3	3	6	6	12
4	4	10	8	20
5	5	15	10	30
6	6	21	12	42
7	7	28	14	56

   Answer (a) is the totals for the thirtieth, not the thirty-first day.
   Answer (c) has the numbers reversed.
   Answer (d) has the wrong number of kernels of corn. The day number is multiplied by 2 and not by 3.

2. Answer: (b) triangle
   The next figure in the pattern will be a circle and have 7 dots. The next 2 figures, a triangle and square, will have 8 and 9 dots, respectively. So, the next circle will have 10 dots. The next circle will have 13 dots. Keep adding 3 to the number of dots in the circle. That gives a circle with 16 dots and then 19 dots. The figure after the circle is a triangle and it will have one more dot than the circle or 20 dots. Therefore, the figure with 20 dots is a triangle.

3. Answer: (c) multiply the first number by 2 and then add 7; 89
   To solve this type of problem, try each of the answers and see if it works on each of the given rows. Answer (a), adding 2 to the previous 2nd number does not involve the 1st number at all. The rule must be used on the 1st number to get the 2nd number. Answer (b) is not correct because $-1+1=0$, which is the next 1st number, not the corresponding 2nd number. Answer (c) is the correct answer—it works for each row and gives a 2nd number of 89 if the 1st number is 41.

   $$2-(1)+7=5, 2(2)+7=11, 2(41)+7=89$$

4. Answer: (a) 6
   First, always try solving the problem with the calculator. In this case the number is too big and the calculator will not give an answer with a digit in the ones place. So, using the calculator, see if a pattern appears. Be sure to try enough numbers that a definite pattern is established.

8^0	1
8^1	8
8^2	64
8^3	512
8^4	4096
8^5	32768
8^6	262144
8^7	2097152
8^8	16777216

   Look at the ones digit in the answers: 8, 4, 2, 6, 8, 4, 2. The pattern in the ones place is 8, 4, 2, 6. So, to find the ones place for the exponent of 12, repeat the pattern until 12 numbers are written. It looks like: 8, 4, 2, 6, 8, 4, 2, 6, 8, 4, 2, 6. There will be a 6 in the ones place for the answer to 8^{12}.

SOLVING ONE-VARIABLE EQUATIONS AND INEQUALITIES ALGEBRAICALLY

In math, a variable is a letter found in an expression, equation, or inequality. Its job is to act like a gift-wrapped present. The gift inside the box is unknown until the gift is opened. The value of the variable is unknown until either a number is identified as the value of the variable, or an equation or inequality is manipulated so that the variable is alone on one side of the equal sign or inequality sign, and a number is alone on the other side of the equal or inequality sign. In the expression $4x + 9$ the variable, x, is a mystery. There is not enough information to determine its value. The meaning of the variable comes from the words in the description of the problem.

Example 1

Evaluate the expression $4x + 9$ when $x = 2$.
   Answer: 17
   Now, there is a value given for x; x can be replaced by the number 2. Write the expression replacing the variable with an empty set of parenthesis.

$$4(\ ) + 9$$

Next, insert the given value of x into the parenthesis and evaluate the expression.

$$4(2) + 9 = 8 + 9 = 17$$

If $x = 0.5$ then the value of the expression will be $4(0.5) + 9 = 2 + 9 = 11$. The value of the expression depends on the given value of the variable, x.

Example 2

Which value of x results in the greatest value for the expression $3|x-2| + 6$?

(a) $x = -5$
(b) $x = 0$
(c) $x = 2$

Answer: (a) $x = -5$
   Insert the given numbers in place of the variable x one at a time and find the value of the expression.

(a) $3|-5-2| + 6 = 3|-7| + 6 = 3(7) + 6 = 21 + 6 = 27$
(b) $3|0-2| + 6 = 3|-2| + 6 = 3(2) + 6 = 6 + 6 = 12$
(c) $3|2-2| + 6 = 3|0| + 6 = 3(0) + 6 = 0 + 6 = 6$

(a) $x = -5$ results in the greatest value for the expression.

An equation is formed when an expression is made equal to something. It has an equal sign with numbers and/or variables on both sides. An equation is a true statement when the expression on the left means the same thing as the expression on the right. Picture an old-fashioned pan balance.

The scale is in balance when both sides weigh the same amount or are equal. The objects on both sides do not need to be identical; they just need to weigh the same total amount. For instance, two 5-pound weights on one side and one 10-pound weight on the other side would have the scale balanced. Both sides are equal even though they are not identical. The same thing is true of the simple equation $x + 6 = 8$. The equal sign says that both sides are equal. The challenge is to find a value for the variable x that keeps the equation true. This is accomplished by following mathematical steps to "solve" the equation—to find the value or values of x that make the equation true.

There are three types of one-variable equations. A *conditional* equation has a single solution. For example, the equation $x + 6 = 8$ is a conditional equation. There is only number, 2, that makes the equation true. An *identity* is an equation that is true for all numbers. For example, $4x + 8 = 4x + 8$ is an identity. The left hand side of the equation and the right hand side of the equation are identical. No matter what number is substituted for x, the equation will always be true. An *inconsistent* equation is an equation for which there is no solution. For example, the equation $x + 1 = x$ has no solution because any number plus 1 cannot equal the original number. Suppose $x = 6$. Substituting 6 into the above equation results in $6 + 1 \neq 6$. Clearly, 7 does not equal 6.

To solve an equation, it is necessary to get the variable (letter, usually x) by itself on one side of the equation and everything else on the other side of the equation. Addition and subtraction are inverse operations. This means that they undo each other. To move a term that is added to the other side of the equation, subtract that term from both sides of the equation. Likewise, to move a term that is subtracted to the other side of the equation, add that term to both sides of the equation. Similarly, multiplication and division are inverse operations. They also undo each other.

The distributive property (formally known as the distributive property of multiplication over addition) is often needed in solving equations. The distributive property can be written as $a(b + c) = ab + ac$.

Likewise, $a(b - c) = ab - ac$.

Example 1

Solve: $x + 12 = 22$.
    Answer: $x = 10$

To solve this equation, it is necessary to get x by itself. It would be by itself if there weren't a $+12$ with it. To undo $+12$, subtract 12 from both sides. Remember, whatever you do to one side of an equation, you must do to the other side of the equation. Subtracting 12 from both sides gives $x + 12 - 12 = 22 - 12$ or $x + 0 = 10$. (It is perfectly okay to write in the $+0$ to help see what is happening in the problem. If it is not needed, skip that step and just write $x = 10$.) It is a good idea to check that the answer obtained is correct. To do a check, plug the value found for the variable into the problem and make sure the two sides of the equation are equal. Ask the question: Does $10 + 12 = 22$? If the answer to the question is yes, the solution is correct. Whenever it is possible to check an answer, do so! Yes, it may take some time, but then you will be assured that the answer is correct.

Example 2

Solve: $2x - 15 = 31$.
    Answer: $x = 23$

This problem is little harder than the one above because it will require an additional step. First, get the $2x$ term by itself. To do that, since 15 is subtracted from both sides of the equation, add it to both sides of the equation. Therefore, $2x - 15 + 15 = 31 + 15$. Now, $2x = 46$. Almost there. The problem is asking for the value of x, not the value of $2x$. Because the x is multiplied by 2, to undo multiplication by 2, it is necessary to divide both sides of the equation by 2, $\frac{2x}{2} = \frac{46}{2}$, which means that $x = 23$. It is left to the reader to do a check.

Example 3

Solve: $5x - 17 = 3x + 23$.
    Answer: $x = 20$

Again, this problem is a little harder than the previous problem because there are terms involving the variable on both sides of the equation. Mathematically, it is fine to move the $5x$ term to the right or the $3x$ term to the left. Either is correct. However, it is suggested to first look at the coefficients and always move the term with the smaller coefficient to the side with the term with the larger coefficient. The reason for this is so that the eventual coefficient will be positive, and it will not be necessary to divide by a negative number. It is easy to make a mistake when dividing by a negative number because of all the rules involving signed numbers.

Because 5 is bigger than 3, subtract $3x$ from both sides of the equation.
This results in $5x - 17 - 3x = 3x + 23 - 3x$ or $2x - 17 = 23$.
Now, to undo -17 from the left hand side, add 17 to both sides of the equation.
Therefore, $2x - 17 + 17 = 23 + 17$.
Thus, $2x = 40$, and dividing both sides by 2 (always find the value for just x, not for some number times x) leaves $x = 20$.

Example 4

Solve: $3(x + 7) = 9(x - 5)$.
   Answer: $x = 11$

The first step in solving this problem is to apply the distributive property. Remember, anything outside of the parenthesis multiplies everything inside of the parenthesis. Using the distributive property, the equation becomes $3x + 21 = 9x - 45$. By inspecting the coefficients, it is preferable to move the $3x$ term to the right hand side of the equation.

   Therefore $3x + 21 \underline{-3x} = 9x - 45 \underline{-3x}$.

   After combining like terms, $21 = 6x - 45$. Add 45 to both sides to undo the -45 on the right hand side of the equation.

   Thus, $66 = 6x$ and dividing both sides by 6 gives the final answer of $x = 11$. Now, do the check!

Example 5

Solve: $\frac{4}{5}x = 20$.
   Answer: $x = 25$

When the coefficient is a fraction, as in the above example, there are two ways to solve the problem. The first way, which is rather clunky, is to divide both sides of the equation by the fraction. Dividing $\frac{4}{5}$ by $\frac{4}{5}$ will result in a coefficient of 1 for the x, which is what is wanted. However, the right hand side will be $\frac{20}{\frac{4}{5}}$. This means 20 divided by $\frac{4}{5}$. Remember, when dividing fractions, keep the first number the same, change the divide to multiply and invert (or use the reciprocal) of the second number. So, this results in $20 \times \frac{5}{4}$ or $\frac{100}{4} = 25$.

   The second way to solve this problem is to multiply both sides of the equation by the reciprocal of the fraction. $\frac{5}{4} \cdot \frac{4}{5}x = 20 \cdot \frac{5}{4}$. Any number (except 0) times its reciprocal always equals 1, so $1x$ (or just x) $= \frac{100}{4} = 25$.

Example 6

Solve: $\frac{1}{2}x + \frac{1}{3} = \frac{2}{9}x + 2$.
   Answer: $x = 6$

When an equation involves many fractions, a good strategy is to clear the fractions from the equation. To clear the fractions means to find a common denominator for all the fractions (it need not be the least common denominator—any common denominator will work) and multiply every term in the equation (even the terms that do not contain fractions) by the common denominator. That will result in an equation equivalent to the original problem, but the equation will not contain any fractions and will be easier and faster to solve.

   In the example, a common denominator for the denominators 2, 3, and 9 is 18 because 18 is divisible by 2, 3, and 9. Multiply every term in the equation by 18.

   $18 \times \frac{1}{2}x + 18 \times \frac{1}{3} = 18 \times \frac{2}{9}x + 18 \times 2$. The original denominators in the problem will all cancel into the common denominator, which is being multiplied throughout. 2 goes into 18 nine times, and $9 \times 1 = 9x$. Continue this process for each term so that the equivalent equation now becomes $9x + 6 = 4x + 36$. Subtracting $4x$ from both sides of the equation yields $5x + 6 = 36$. Now, subtract 6 from both sides of the equation so that $5x = 30$. Lastly, divide both sides by 5 to determine that $x = 6$.

Example 7

Solve: $.01x + .003 = 2x + .07$.
  Answer: $x = -\frac{67}{1990}$

Solving an equation with decimals can be challenging. A helpful strategy is to get rid of the decimals and rewrite the equation as an equivalent equation without any decimal points. To do this, determine which is the smallest decimal in the equation. (The smallest decimal will be the one with the most places after the decimal point. In other words, if the decimals in the equation were 0.03, 0.004, and 0.00005, 0.00005 would be the smallest decimal because it has 5 places after the decimal point.) In the example, the smallest decimal is 0.003. Now, move the decimal point of every term in the equation (including terms that have no visible decimal point) three places to the right. It is helpful to add a decimal point, when needed, and to add zeros following the last decimal place to make sure that all the terms have, in this case, three decimal places.

Rewrite the equation to be $0.010x + 0.003 = 2.000x + 0.070$. Now, move all the decimal points three places to the right (this is the same as multiplying each term by 1,000) so that the equivalent equation becomes $10x + 3 = 2000x + 70$. Subtracting $10x$ from both sides leaves $3 = 1990x + 70$. Subtracting 70 from both sides yields $-67 = 1990x$. Divide both sides by 1990 to get $x = -\frac{67}{1990}$. Yes, sometimes answers are fractions. Don't immediately think an error has been made because the final answer is a fraction or decimal. In the real world, some correct answers are fractions or decimals.

Practice Problems

1. Solve: $5(x - 3) = -15 + 5x$.
   (a) 0
   (b) 3
   (c) there is no solution
   (d) all real numbers
2. Solve: $5x + 32 - 3x = 4x + 7x - 4$.
   (a) 9
   (b) 4
   (c) −9
   (d) −4
3. Solve: $3x + 6 = 3(x - 7)$.
   (a) $-\frac{13}{6}$
   (b) there is no solution
   (c) all real numbers
   (d) $-\frac{14}{3}$
4. Solve: $\frac{1}{4}x - 4 = \frac{2}{3}x + \frac{1}{6}$.
   (a) all real numbers
   (b) there is no solution
   (c) 10
   (d) −10

Detailed answer key for the above problems:

1. Answer: (d) All real numbers
   This is an identity. After distributing, the left hand side becomes $5x - 15$. Rewrite the right hand side to be $5x - 15$. Both sides are identical, so all real numbers will be solutions to the equation.
2. Answer: (b) 4
   To solve this equation, it is necessary to first combine the like terms that appear on the same side of the equal sign. On the left hand side, $5x - 3x$ can be combined to make $2x$, while on the right-hand side, $4x + 7x$ can be combined to make $11x$. Therefore, $2x + 32 = 11x - 4$. Subtract $2x$ from both sides, giving $32 = 9x - 4$. Now, add 4 to both sides so that $36 = 9x$; divide both sides by 9 so that $x = 4$.
3. Answer: (b) There is no solution.
   This is an example of an inconsistent equation. After distributing 3 on the right-hand side, the problem becomes $3x + 6 = 3x - 21$. Subtracting $3x$ from both sides gives the false statement that $6 = -21$. Anytime a false statement results, there is no solution.
4. Answer: (d) -10
   Start by clearing the fractions. Find a common denominator for the denominators given in the problem: 4, 3, and 6. The least common denominator (remember, any common denominator will work—it does not have to be the least common denominator) is 12. Now, multiply every term in the equation (including terms that do not include fractions) by 12. This gives $3x - 48 = 8x + 2$. Subtracting $3x$ from both sides results in $-48 = 5x + 2$. Now, subtract 2 from both sides so that $-50 = 5x$ and then divide both sides by 5: $-10 = x$.

USING THE CONCEPTS OF VARIABLE, EQUALITY, AND EQUATION TO GENERATE, INTERPRET, AND EVALUATE ALGEBRAIC EXPRESSIONS

While solving an equation can be challenging, the bigger challenge is reading a word problem and translating those words into a mathematical equation that can then be solved. This chart can be used to translate word phrases into algebraic expressions.

Addition (+)	Subtraction (−)	Multiplication (×)	Division (÷)
add	subtract	multiply	divide
added to	subtracted from	multiplied by	quotient
sum	difference	product	divided by
total	minus	times	
plus	less than	of	
more than	decreased by		
increased by	take away		

Example 1

Translate the following written expression into an algebraic expression:

$$\text{Seven less than a number}$$

Answer: $x-7$

Be careful with this expression. Whatever follows "less than" in a written expression will come first in the subtraction problem.

Example 2

Translate the following written expression into an algebraic expression:

$$\text{Five more than three times some number}$$

Answer: $3x+5$

Example 3

Translate the following written expression into an algebraic expression:

$$\text{Half of some number}$$

Answer: $\frac{1}{2}x$

When translating a word problem to an equation, the first step is to define what the variable will represent. The other information in the problem will then work around that variable.

Example 4

Jason's brother is 6 years older than Jason. Together, their ages total 48. How old are Jason and his brother?

Answer: Jason is 21 and Jason's brother is 27.

This problem has two unknowns, Jason's age and his brother's age. Nothing is known about Jason's age. It is known that his brother is six years older than Jason. So, the variable will represent Jason's age—the person that nothing is known about. His brother's age can be represented with the expression $x + 6$. So far this is what can be written:

$$\text{Jason's age: } x$$

$$\text{Jason's brother's age: } x + 6$$

The next statement in the problem is that their ages total 48.
The translation of this statement is: Jason's age + Jason's brother's age = 48.

Now, substitute the expressions for the words: $x + x + 6 = 48$.

Solve the equation by first combining like terms: $2x + 6 = 48$.
Subtract 6 from both sides: $2x = 42$.
Divide both sides by 2: $x = 21$.

Now substitute the value of x back into the two expressions defining the age of both boys: Jason's age is 21 and his brother's age is $21 + 6$, which is 27. Together their ages should total 48: $21 + 27 = 48$.

Practice Test Questions on Using the Concepts of Variable, Equality, and Equation to Generate, Interpret, and Evaluate Algebraic Expressions

1. Last summer Andy painted 35 more houses than Jason did. They painted 121 houses altogether. How many houses did Andy paint?
   (a) 78 houses
   (b) 43 houses
   (c) 86 houses
   (d) 156 houses
2. Cathy used to run a 9.5-minute mile. Now, she runs a mile in 11.25 minutes. How much longer will it take to run three miles at her new, slower pace than it took her to run three miles at her faster pace?
   (a) $1\frac{3}{4}$ minutes
   (b) $3\frac{3}{4}$ minutes
   (c) $5\frac{1}{4}$ minutes
   (d) $33\frac{3}{4}$ minutes
3. Fifteen less than the quotient of a number and 9 is -4. What is the number?
   (a) 171
   (b) 99
   (c) -171
   (d) $\frac{11}{9}$
4. The sum of three consecutive even integers is 132. What are the integers?
   (a) 42
   (b) 64, 66, 68
   (c) 42, 44, 46
   (d) 43, 44, 45

Detailed answer key for the above problems:

1. Answer: (a) 78 houses
   Andy painted 35 more houses than Jason did. Nothing is known about the number of houses Jason painted.

Jason's houses: h

Andy's houses: $h + 35$

Jason's houses + Andy's houses = 121

$h + h + 35 = 121$

$2h + 35 = 121$

$2h = 86$

$h = 43$

Remember, 43 is the number of houses that Jason painted. The question is asking how many houses Andy painted. So, substitute 43 for h in the expression for the number of houses that Andy painted: $43 + 35 = 78$.

2. Answer: (c) $5\frac{1}{4}$ minutes

   First, find the difference in the amount of time it takes Cathy to run a mile by subtracting 9.5 from 11.25. The difference is 1.75 or $1\frac{3}{4}$. That is how much longer it takes to run 1 mile. The question asks about the time to run 3 miles. So, multiply $1.75 \times 3 = 5.25$ or $5\frac{1}{4}$ minutes.

3. Answer: (b) 99

   The quotient of a number and 9 is written as $\frac{n}{9}$. Fifteen less than that quotient is written as $\frac{n}{9} - 15$. That expression equals -4, so $\frac{n}{9} - 15 = -4$ is the equation that needs to be solved. Start by adding 15 to both sides of the equation. Then, multiply both sides of the equation by 9 giving $n = 99$.

4. Answer: (c) 42, 44, 46

   This problem can be solved by checking each of the 4 answers to see if they meet the requirements of the problem. The question asks for all 3 integers so answer (a) cannot be correct. The sum of the 3 even integers means that the three numbers need to be added together. Only answers (c) and (d) add up to 132. The question is looking for 3 consecutive even integers, so that eliminates answer (d). Therefore, (c) is the correct answer.

   To actually solve this problem, define x as the first even integer. The next even integer would then be $x + 2$ and the third even integer will be $x + 4$. To add the three integers together would be the expression $x + x + 2 + x + 4$ and that expression is equal to 132.

   $x + x + 2 + x + 4 = 132$. Combine like terms, $3x + 6 = 132$, and solve for x:
   $x = 42$; the next integer is $x + 2$ so it is 44 and the last integer is $x + 4$, which is 46.

MANIPULATING ALGEBRAIC EXPRESSIONS AND SOLVING EQUATIONS USING A VARIETY OF TECHNIQUES

Just like pulling on your socks before putting on your shoes is the correct order when getting dressed, multiplying before subtracting is the correct order when doing math calculations. PEMDAS (Please Excuse My Dear Aunt Sally) is the acronym used to remember the order of operations. This stands for:

Parenthesis
Exponents
Multiply/**D**ivide in the order they are presented in the problem, moving from left to right
Add/**S**ubtract in the order they are presented in the problem, moving from left to right

Example 1

$$\frac{2+3}{11-1} \times 20$$

The numerator and denominator of a fraction are treated as though they are within a set of parenthesis. Therefore, $2 + 3 = 5$ and $11 - 1 = 10$ are calculated first, giving $\frac{5}{10} \times 20$.

Then, the division comes before the multiplication so it is performed first: $0.5 \times 20 = 10$.

Example 2

$$1000 \div 5 \times 3 - 40 + (2 \times 3^2)$$

Answer: 578

(2×3^2)	Complete what is in the parenthesis first.
$(2 \times 9) = 18$	
$1000 \div 5 \times 3 - 40 + 18$	Divide first.
$200 \times 3 - 40 + 18$	Then multiply.
$600 - 40 + 18$	Then subtract.
$560 + 18$	Finish by adding.

The answer is 578.

SIMPLIFYING OR COMBINING LIKE TERMS

Simplifying or combining like terms is used frequently in solving equations. A term is made up of a variable with or without an exponent, a single number, or both of these multiplied together. Terms are separated by + or − signs. These are examples of terms: $4xy$, $5z^3$, y, 10. The expression $2a + 6ab^2 + 6$ is made up of three terms. Terms can be combined or put together if the variable and the exponent match.

Example 1

Simplify $2a + 3a^2b + 6b + 9a + 4a^2b + 5b^2$.
  Answer: $11a + 7a^2b + 6b + 5b^2$
  $2a$ and $9a$ can be combined. $3a^2b$ and $4a^2b$ can be combined.
  So, this expression is simplified to become $11a + 7a^2b + 6b + 5b^2$. None of the other terms can be combined.

Example 2

Simplify $2xy + 3x^2y - (xy - 5x^2y)$.
  Answer: $xy + 8x^2y$
  First, distribute the negative sign to the terms inside the parenthesis.

$$2xy + 3x^2y - xy + 5x^2y$$

Then combine the like terms.

$$2xy - xy + 3x^2y + 5x^2y = xy + 8x^2y$$

BASIC DISTRIBUTIVE PROPERTY

Just like the basic distributive property, this same property can be used to multiply binomials—expressions with two terms. The acronym FOIL is commonly used in this process but it really is just another use of the distributive property. FOIL stands for first, outer, inner, last, which is the order in which the terms are multiplied.

Example 1

$$(x + 3)(x - 5)$$

  Answer: $x^2 - 2x - 15$
  Start by multiplying the first term in each binomial: $x \times x = x^2$.
  Then multiply the outer terms: $x \times -5 = -5x$. The x has now been distributed to both terms in the second binomial.
  Now multiply the inner terms: $3 \times x = 3x$. Then multiply the last terms together: $3 \times -5 = -15$.
  The 3 has now been distributed to both terms in the second binomial. Put all of the pieces together: $x^2 - 5x + 3x - 15$. Then combine the like terms to get the final answer: $x^2 - 2x - 15$.
  Be careful not to lose any of the negative signs.

Example 2

$$(x + 9)(x - 9)$$

Answer: $x^2 - 81$

$$(x + 9)(x - 9) = x^2 - 9x + 9x - 81 = x^2 - 81$$

This is an example of the difference of two squares because x^2 and 81 are both perfect squares.

FACTORING AN EXPRESSION

Factoring an expression is the reverse of the distributive property. When looking at a polynomial, an expression with more than one term, anything that all the terms have in common is factored or divided out of the expression and written in front of the parenthesis containing the rest of the expression. The last step in factoring should be to use the distributive property and be sure that the original expression is the answer.

Example 1

Factor $4y + 6$.
   Answer: $2(2y + 3)$
   Write the expression as the product of its factors: $(2 \times 2 \times y) + (2 \times 3)$. Each term has a factor of 2 so divide each term by 2. The 2 is written in front of the parenthesis and the factors remaining after the term is divided by 2 are written inside the parenthesis. The answer is $2(2y + 3)$.
   Another way of looking at this problem is to look at $4y$ and 6 and ask what these terms can be evenly divided by. Both the 4 and the 6 can be divided by 2. $4y \div 2 = 2y$ and $6 \div 2 = 3$. The 2 goes outside the parenthesis and the $2y$ and the 3 go inside the parenthesis along with the $+$ sign that separated them.
   Check the answer by using the distributive property, $2(2y + 3) = 2 \times 2y + 2 \times 3 = 4y + 6$, which is the original expression, so the factoring was correct.

Example 2

Factor $9z^3 + 3z^2 + 6z$.
   Answer: $3z(3z^2 + z + 2)$

$$\text{③} \times 3 \times \text{ⓩ} \times z \times z + \text{③} \times \text{ⓩ} \times z + 2 \times \text{③} \times \text{ⓩ}$$

Each term has a 3 and a z in common. These are factored out, which results in this expression: $3z(3z^2 + z + 2)$.
   A special type of factoring is found with a quadratic trinomial such as $x^2 + 5x + 6$. This is an expression with three terms where the first term has a variable with an exponent of 2. When this expression is factored, it will be written as $(x + 2)(x + 3)$.

Example 3

Factor $x^2 + 5x + 6$.

Answer: $(x + 2)(x + 3)$

The first step to make this happen is to list all the pairs of factors that when multiplied together will give a product of 6 (the constant at the end of the expression). These same pairs of factors when added together need to give a sum of 5 (the coefficient of the middle term). Because 6 is a positive answer, both factors will be positive or both factors will be negative. Because 5 is a positive number, both factors will have to be positive numbers.

$$\text{Factors of 6}$$
$$1 \times 6$$
$$2 \times 3$$

Because the factors of 2 and 3 give a product of 6 and a sum of 5, those are the numbers needed inside the parenthesis. The first term, x^2, means that each of the two factors will start with an x. So, the two factors start like this $(x \quad)(x \quad)$. Then, the factors of 6 are put into the parenthesis. Both factors are positive numbers so they are attached to the variable x with a plus sign.

$(x + 2)(x + 3)$ is the answer and if the distributive property is used then the original expression of $x^2 + 5x + 6$ is the result.

Example 4

Factor $x^2 - x - 20$.

Answer: $(x + 4)(x - 5)$

Start with $(x \quad)(x \quad)$.

Because 20 is a negative number, one of the factors must be negative. Find the factors of -20 that added together will give a sum of -1 (the coefficient of the middle term).

Factors of 20	Add and Equal -1
-1×20	19
1×-20	-19
-2×10	8
2×-10	-8
-4×5	1
4×-5	-1

In the table above shows the correct factors will be 4 and −5. So the expression will be factored as $(x + 4)(x − 5)$. Checking it with FOIL results in $x^2 − 5x + 4x − 20 = x^2 − x − 20$. This is the original expression, so the answer is correct.

Sometimes the first term will also have a coefficient. That makes factoring a little more complicated, but there is a method to help. The first step is to see if a common factor can be divided out of all three terms.

Example 1

Factor $2x^2 − 22x + 56$.

Answer: $2(x − 4)(x − 7)$

A factor of 2 can be divided out of each of the terms, giving $2(x^2 − 11x + 28)$. The factors of +28 that add to −11 are −4 and −7. The factored expression is $2(x − 4)(x − 7)$.

It is not always possible to factor out a common number. In that case, using a factor box may help.

Example 2

Factor $2x^2 − x − 6$.

Answer: $(2x + 3)(x − 2)$.

To factor this expression the first method in this section will not work because there is a coefficient, or number, in front of the x^2 term. So, the next step is to see if the 2 can be factored out of all three terms. A 2 cannot be factored out of the middle term. So, a third method is needed.

Create a box and write the first term in the top left box and write the last term in the bottom right box.

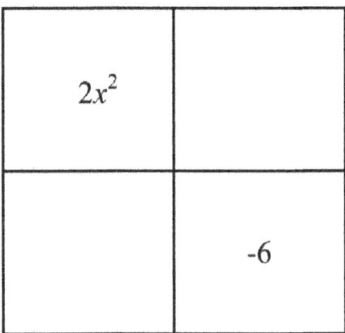

Then, multiply the coefficient of the first term by the number in the last term—in this example that would be $2 \times -6 = -12$. Find factors of −12 that when added together give the coefficient of the middle term—which will be −1 in this example. These factors will be −4 and 3. Fill in the other two boxes using these factors and the variable x. It doesn't matter which number goes in which box.

$2x^2$	$-4x$
$+3x$	-6

Now, factor each row and each column. Always use the sign of the box closest to the top or closest to the left side of the table. The variable x can be factored out of the first column, $2x^2$ and $3x$. Because the top left box is positive, the x will be positive. The number 2 can be factored out of the second column, $-4x$ and -6. Because the top right box is negative, the number factored out will be -2. Working across the rows, $2x$ can be factored out of the first row. The top left box is positive, so the $2x$ will be positive. Finally, factor a 3 out of the bottom row. Because the bottom left box is positive, the 3 will be positive.

Now the table will look like this:

	x	-2
$2x$	$2x^2$	$-4x$
$+3$	$+3x$	-6

Take the terms on the outside of the box and put them together to form the two binomials. They will be $(2x + 3)(x - 2)$. Check this by multiplying these two factors together using FOIL: $2x^2 - 4x + 3x - 6 = 2x^2 - x - 6$, which is the trinomial that was being factored. The solution is correct.

It is possible that none of these three methods will provide an answer, such as the trinomial $x^2 + 5x - 7$. Then the quadratic formula must be used. But these types of problems are not represented on the test and will not be addressed in this book.

If a quadratic trinomial is equal to 0 and the trinomial can be factored into two or more factors, the equation can be solved for the value or values of x. This is done by using the zero product property of multiplication, which says that when two numbers are multiplied together to give a product of zero, one or both of the numbers must be zero. So, all of the factors will be set equal to 0 and solved to find the value of x.

Example 3

Solve $2x^2 - x - 6 = 0$.
   Answer: The two solutions for x are 2 and $-\frac{3}{2}$
   First, factor the trinomial—this was shown in example 2.

$$(2x + 3)(x - 2) = 0.$$

Then, solve for x.

$$2x + 3 = 0 \qquad x - 2 = 0$$

$$2x = -3 \qquad x = 2$$

$$x = -\frac{3}{2}$$

The two solutions for x are 2 and $-\frac{3}{2}$. There will be two solutions because this equation is not linear (doesn't make a line). It is the equation for a parabola that will cross the x-axis in two places, thus the two values of x. This topic will be covered in more depth in chapter 4.

Practice Test Questions Using the Concepts of Manipulating Algebraic Expressions and Solving Equations Using a Variety of Techniques

1. Simplify this expression by combining like terms: $4a^2b - 6ab + 9ab^2 - a^2b + ab - 4$.
   (a) $3a^2b - 5ab + 9ab^2 - 4$
   (b) $7a^2b - 4$
   (c) $3a^4b^2 - 6a^2b^2 + 9ab^2 - 4$
   (d) $7a^2b + 5ab + 9ab^2$
2. Factor the expression $x^2 - 28x + 196$.
   (a) $(x + 14)(x - 14)$
   (b) $(x + 14)^2$
   (c) $(x - 14)^2$
   (d) $(x - 14)(x + 14)$
3. Connie gives Wendy a challenge. She says to take a number and add three to it. Then, take the same number and subtract seven from it. When these two quantities are multiplied together, they equal 0. What positive number did Wendy choose?
   (a) −3
   (b) 7
   (c) 4
   (d) 0

4. The area of a garden is 24 ft². The length of the garden is 5 feet more than the width of the garden. Which of these measurements could be the values for the length and width of the garden?
   (a) 3, 8
   (b) 3, −8
   (c) −3, −8
   (d) 4, 6

Detailed answer key for the above problems:

1. Answer: (a) $3a^2b - 5ab + 9ab^2 - 4$
   The first and the third term are combined and the second and fifth term can be combined. Answers (b) and (d) are incorrect because the exponents in the terms with the variables are not all the same so the numbers cannot all be combined. Answer (c) is wrong because when combining like terms, the exponents will not change.
2. Answer: (c) $(x - 14)^2$
   Answers (a) and (d) are the same answer. There cannot be two correct answers so both of them can be eliminated. Working backward from the answers, (b) would be $(x + 14)(x + 14) = x^2 + 28x + 196$. Answer (c) is $(x - 14)(x - 14) = x^2 - 28x + 196$.
3. Answer: (b) 7
   The first quantity is $(x+3)$. The second quantity is $(x - 7)$. The equation would then be $(x + 3)(x - 7) = 0$. Each factor is set equal to 0. $x + 3 = 0$ and $x - 7 = 0$—so the values of x are −3 and 7. Wendy chose a positive number so her number is 7.
4. Answer: (a) 3, 8
   The width is the unknown measurement so it will be represented by w. The length is 5 more than the width, which can be written as $w + 5$. Area is found by multiplying the length and the width, so the area is $w(w + 5)$. The problem states that the area is 24 square feet, so $w(w + 5) = 24$. Using the distributive property, $w^2 + 5w = 24$. Then subtract 24 from both sides so that the trinomial can be factored: $w^2 + 5w - 24 = 0$. The factors of −24 that add up to 5 are −3 and 8. So the factored equation is $(w - 3)(w + 8) = 0$. Next, $w - 3 = 0$ and $w + 8 = 0$. The values of w are 3 and −8. A length cannot be negative so the width must be 3 feet. The length is 5 feet more than the width so the length is 8 feet. The area is then $3 \times 8 = 24$, which is the correct area.

APPLYING ALGEBRAIC PRINCIPLES TO REPRESENT AND SOLVE WORD PROBLEMS INVOLVING RATIOS AND PROPORTIONS

A *ratio* is a fraction that compares two quantities or numbers. The ratio of 10 dogs to 3 dog trainers can be written as $\frac{10}{3}$ or 10:3. The order that the numbers appear in the ratio is important. The words before the word *to* are the top of the fraction and the words after the word *to* are the bottom of the fraction. The fraction should always be simplified.

Example 1

There are 14 lifeguards on a very crowded beach on the 4th of July. It is estimated that there are 400 people on the beach. What is the ratio of people to lifeguards?
Answer: $\frac{200}{7}$
The word before the word *to* is *people*. The word after the word *to* is *lifeguards*. There are 400 people and 14 lifeguards, so the ratio is 400:14 or $\frac{400}{14}$. This needs to be simplified by dividing each number by 2 giving $\frac{200}{7}$. The ratio of people to lifeguards is $\frac{200}{7}$.

Example 2

The Devanney Demons had a winning season in football. They won 15 games and only lost 4.
  (a) What is the ratio of games won to games lost?
     Answer: The ratio of games won to games lost is $\frac{15}{4}$.
  (b) What is the ratio of games won to total games?
     Answer: The ratio of games won to total games is $\frac{15}{19}$.

The denominator, 19, was found by adding the games won to the games lost to get a total of 19 games played.

A *proportion* is an equation with two ratios that are equal. This equality can be checked by using a method called cross-multiplication. The denominator of one fraction is multiplied by the numerator of the other fraction—across the equal sign. Then the other denominator is multiplied by the other numerator. If both products are equal, then the ratios are proportional.

Example 1

Are these ratios proportional? $\frac{8}{12} = \frac{6}{9}$
  Check by multiplying $9 \times 8 = 12 \times 6$
  $72 = 72$, so these ratios are proportional

Example 2

Are these ratios proportional? $\frac{7}{8} = \frac{3}{4}$
  Answer: No, the ratios are not proportional.
  Check by multiplying $8 \times 3 = 4 \times 7$
  $24 \neq 28$ so these ratios are not proportional

The cross-multiplication process can be used to solve a proportion.

Example 3

$$\frac{3}{4} = \frac{x}{22}$$

Answer: $x = 16.5$

Cross-multiplication leads to this equation: $4x = 22 \times 3$
$4x = 66$
Divide both sides by 4 $x = 16.5$

Example 4

$$\frac{2\frac{1}{2}}{5} = \frac{y}{3\frac{1}{5}}$$

Answer: $y = 1\frac{3}{5}$

$$5y = \frac{5}{2} \times \frac{16}{5}$$

$$5y = 8$$

$$y = \frac{8}{5} = 1\frac{3}{5}$$

Proportions are used when making scale drawings of a house or a model of something, for example. They can also be used when increasing or decreasing the amount of people served by a recipe. Another use comes in comparing similar triangles.

When solving a word problem involving proportions, it is important to define the ratio or define the quantity that will be in the numerator and then define the quantity that will be in the denominator.

Example 1

The science club is hosting a science activity day for the local elementary school. They have 60 minutes for each class to build a volcano. They have determined it will take 8 pounds of flour to make 3 volcanoes. How many pounds of flour will be needed if 11 classes attend and each class builds 2 volcanoes?

Answer: $58\frac{2}{3}$ pounds of flour

First, the number of volcanoes to be built must be calculated. Eleven classes each making 2 volcanoes will be 22 volcanoes altogether. The ratio that is needed to solve this problem will be $\frac{\text{pounds of flour}}{\text{number of volcanos}}$ or $\frac{\text{number of volcanos}}{\text{pounds of flour}}$. The order does not matter but it must remain consistent throughout the problem. Using the first ratio, the proportion will look like this:

$$\frac{\text{pounds of flour}}{\text{number of volcanos}} = \frac{8}{3} = \frac{f}{22}$$

Read this proportion as: 8 pounds of flour is to 3 volcanoes as I-don't-know-how-many pounds of flour is to 22 volcanoes. Then use cross-multiplication to solve the proportion. $3f=176, f=58\frac{2}{3}$. The science club will need $58\frac{2}{3}$ pounds of flour for the build a volcano activity. Notice that the 60 minutes given in the problem has nothing to do with finding the answer to the question.

Example 2

Rational High School is adding a new swimming pool wing to their current building. The architect brings the plans to the student council meeting to show the student. On the plans, the pool measures 4.1 inches by 8.2 inches. The scale on the drawing is 1 inch = 20 feet. What are the actual measurements of the pool?

Answer: The pool is 164 feet long

This is actually two problems. First, a proportion is needed to find the actual width. $\frac{inches}{feet} = \frac{1}{20} = \frac{4.1}{x}$, $x = 20 \times 4.1 = 82$. The pool is 82 feet wide.
A second proportion is needed to find the actual length. $\frac{inches}{feet} = \frac{1}{20} = \frac{8.2}{x}$,

$x = 20 \times 8.2 = 164$. The pool is 164 feet long.

Practice Test Questions Using Algebraic Principles to Represent and Solve Word Problems Involving Ratios and Proportions

1. What is the ratio of squares to stars?
    (a) $\frac{2}{8}$
    (b) $\frac{2}{4}$
    (c) $\frac{1}{2}$
    (d) $\frac{2}{1}$

2. A group of 50 students were selected to wear pink ribbons to an assembly. During the assembly, their principal selected 35 random students to pose for a spirit picture. She notices that 14 of these students are wearing the pink ribbons. How many students were at the assembly?
    (a) 50
    (b) 490
    (c) 125
    (d) not enough information given

3. Katie can solve 17 math problems in 23 minutes. How many minutes will it take her to complete 11 problems, to the nearest minute?
    (a) 15
    (b) 14.88
    (c) 14
    (d) 8

4. A party is being held to celebrate the last day of finals. Darleen has decided to serve hot dogs, chips, and punch. The recipe for the punch has several ingredients. To serve 30 people, the recipe calls for $\frac{1}{2}$ gallon of orange sherbet, a 6-ounce can of frozen orange juice and a 2-liter bottle of ginger ale. How much of each ingredient will be needed to serve 100 people?
   (a) 6000 gallons, 500 ounces, 1500 liters
   (b) $1\frac{2}{3}$ gallons, $\frac{1}{3}$ ounce, 4 liters
   (c) $6\frac{2}{3}$ gallons, 20 ounces, $1\frac{2}{3}$ liters
   (d) $1\frac{2}{3}$ gallons, 20 ounces, $6\frac{2}{3}$ liters

Detailed answer key for the above problems:

1. Answer: (c) $\frac{1}{2}$
   There are two squares and four stars. The ratio is squares to stars so the number of squares is the numerator and the number of stars is the denominator. Then, the fraction must be simplified $\frac{2}{4}=\frac{1}{2}$.
2. Answer: (c) 125
   Set up the ratios like this $\frac{\text{students wearing ribbons}}{\text{total students}}$. Of the 35 total students selected for the picture, 14 are wearing ribbons. So, the first half of the proportion will be $\frac{14}{35}$. Before the assembly started, 50 students out of all the students going to the assembly were wearing ribbons. So the second half of the proportion will be $\frac{50}{s}$. Set these two ratios equal to each other and solve for s. $\frac{14}{35} = \frac{50}{s}$. $14s=1750$. $s=125$.
3. Answer: (a) 15
   Set up the ratios with the number of math problems in the numerator and the number of minutes in the denominator. $\frac{17}{23}=\frac{11}{m}$. Then, cross-multiply: $17m=253$. $m=14.88235294$. The answer asked for the number of minutes, rounded to the nearest minute. That means to round to the nearest whole number, which will be 15.
4. Answer: (d) $1\frac{2}{3}$ gallons, 20 ounces, $6\frac{2}{3}$ liters
   Three different proportions are needed to solve this problem. The first will be used to find the amount of orange sherbet needed. Use the ratio of $\frac{\text{amount of ingredient}}{\text{number of people}}$. The recipe always gives the first half of the proportion and the question is asking for the numerator in the second half of the proportion. To find the amount of orange sherbet the proportion will be $\frac{1/2}{30} = \frac{x}{100}$. This finds that $1\frac{2}{3}$ gallons of orange sherbet will be needed. The orange juice is found by the proportion $\frac{6}{30} = \frac{x}{100}$. Twenty ounces of orange juice will be needed. Answer (b) used the proportion $\frac{1}{30} = \frac{x}{100}$. This would find the number of cans needed, $3\frac{1}{3}$ cans. The units don't match the answer given. The proportion for the number of liters of ginger ale, not the number of bottles of ginger ale, is $\frac{2}{30} = \frac{x}{100}$. This finds that $6\frac{2}{3}$ liters of ginger ale are needed.

Chapter Four

Algebra

GRAPHING LINES AND INEQUALITIES

The slope of the line containing two points is the measure of the steepness of the line. Slopes may be positive (the line is going uphill), negative (the line is going downhill), zero (the line is horizontal), or undefined (the line is vertical). Slope is always represented by the letter m, and equals rise over run, or the difference of the y-values divided by the difference of the x-values.

Example 1

Find the slope of a line containing the points (1,2) and (5,5).
 Answer: $\frac{3}{4}$
 The slope is found by letting the point (1,2) = (x_1, y_1) and the point (5,5)=(x_2, y_2). Remember, x_1 must go in the same ordered pair as y_1 and x_2 must go in the same ordered pair as y_2. Also, make sure to distinguish between subscripts (or numbering devices—the numbers that follow x and y in the above example) and exponents (which are written as superscripts or numbers above the letters). Nothing is squared in the slope formula. Plug these values into the formula:

$$m = \frac{y_2 - y_1}{x_2 - x_1} \text{ yields } \frac{5-2}{5-1} \text{ or } \frac{3}{4}.$$

To go from one point to the next point on the line, go up three units and to the right four units.
 If a slope is negative, go down the number of units in the numerator of the fraction, and to the right the number of units in the denominator. Always go to the right when going from point to point.
 A linear equation takes the form of $Ax+By=C$ and always graphs into a straight line. There are several ways to graph a line. If possible, it is best to have integer values for x and y so that the graph is precise. One method, which sometimes does not result in integer values, is to select three values for x and find their corresponding y values. Technically, a line is determined by two points, but it is a good idea to have a third point as a check.

Example 2

Graph the line $y=5x+3$ by selecting values. Pick x values of -1, 0 and 1.
Answer:

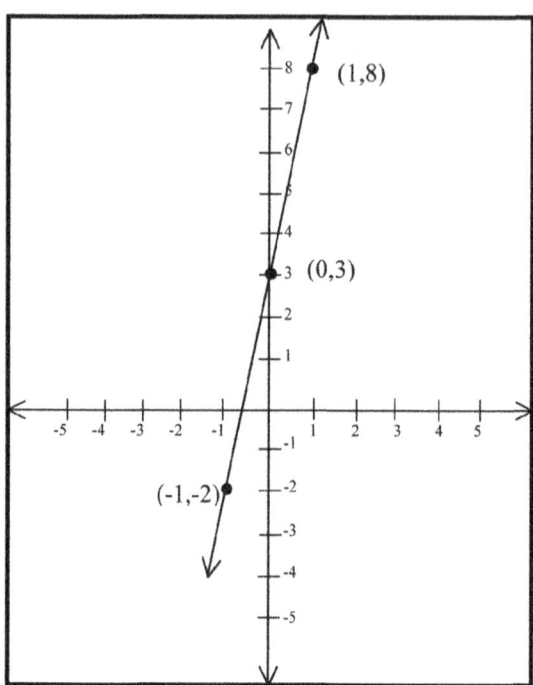

To solve this problem, substitute each value for x into the equation, and solve for the corresponding value of y. When $x=-1$, $y=5(-1)+3=-5+3=-2$. Therefore, the ordered pair $(-1,-2)$ is on the line. Next, let $x=0$ and solve for y. When $x=0$, $y=5(0)+3=0+3=3$. Plot the next point, $(0,3)$, on the line. Finally, let $x=1$ so that $y=5(1)+3=5+3=8$. Finish the line by plotting the point $(1,8)$. Note that the line is going uphill. From this, we can determine that the slope is positive. Start at the point $(0,3)$. To get to the next point, notice that it is necessary to go up five units and over to the right one unit. This means that slope of the line is $\frac{5}{1}$ or 5. Also notice that the line crosses the y-axis at the point $(0,3)$. This point is called the y-intercept, as explained below.

The second method of graphing lines is to use intercepts. The y-intercept of a line is where the line crosses the y-axis, while the x-intercept of the line is where the line crosses the x-axis. To find the y-intercept, let x equal 0 and solve for y. To find the x-intercept of a line, let y equal 0 and solve for x. Not all lines, however, have x and y intercepts. Horizontal lines, which either equal the x-axis (when $y=0$) or are parallel to the x-axis, have no x-intercepts. These lines always take the form of $y =$ some number, such as $y=14$. Likewise, vertical lines, which either equal the y-axis (when $x=0$) or are parallel to the y-axis, have no y-intercepts. These lines always take the form of $x =$ some number, such as $x=-3$.

Example 3

Graph $3x+4y=12$ using intercepts.
Answer:

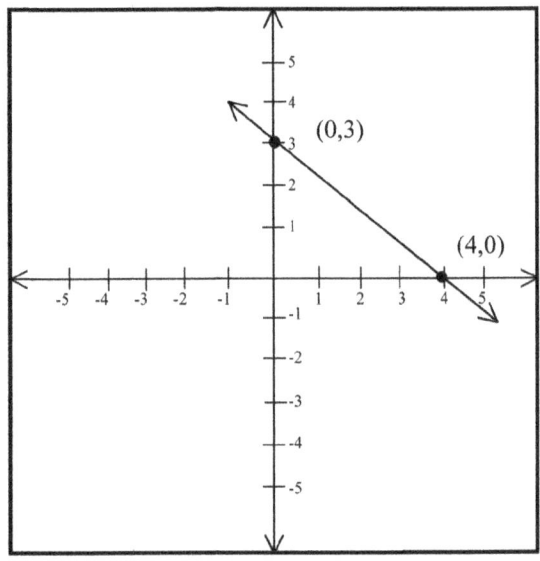

To find the x-intercept, let $y=0$ and solve for x. $3x+4(0)=12$, or $3x=12$. Dividing both sides by 3, $x=4$. So the point (4,0) is on the line. Likewise, to find the y-intercept, let $x=0$ and solve for y. $3(0)+4y=12$, or $4y=12$. Dividing both sides by 4, $y=3$. So the point (0,3) is on the line.

Example 4

Graph the line $y=4$ and determine its slope.
Answer:

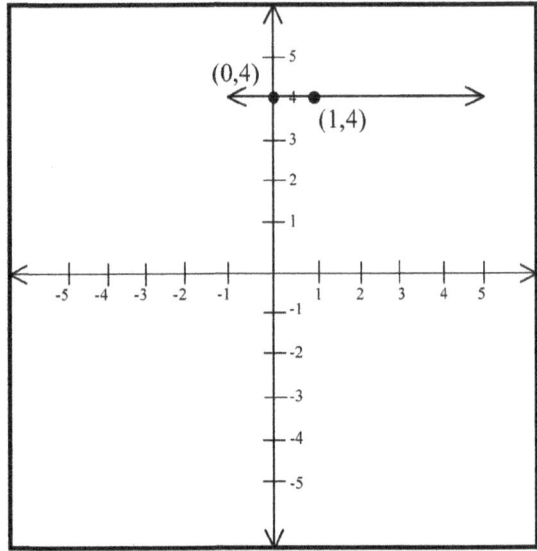

Remember, not all lines can be graphed using intercepts. Whenever the line is a form that $y=$ some number (and there is no x in the equation of the line), the line is a horizontal line. It does not matter what value x takes on; y is always 4. Thus, the points (0,4) and (1,4) are on the line. To determine the slope of this line, use the slope formula $m=\frac{y_2-y_1}{x_2-x_1}$, where the point (0,4) = (x_1,y_1) and the point (1,4) = (x_2,y_2). Plugging these values into the formula $m=\frac{y_2-y_1}{x_2-x_1}$ yields $\frac{4-4}{1-0}=\frac{0}{4}=0$. Slopes of horizontal lines (those of the form $y=$ some number) always have a slope of 0. The easy way to remember this is to associate the letter z in the word *horizontal* with the letter z in the word *zero*.

The third method of graphing a line is by inspection, which means to put the equation into the form $y=mx+b$, where m is the slope of the line and b is the y-intercept of the line. Once the line is in that form, start at the point $(0,b)$ or the y-intercept, and move to the next point by using the slope. For example, if the y-intercept equals 6 and the slope $=\frac{-3}{2}$, graph the point (0,6) and then go down three units (down because the slope is negative) and over to the right two units (always go the right) and graph the next point on the line. Continue this process again to find a third check point on the line and connect the points to draw the line.

Example 5

Graph $3x+4y=12$ by inspection.
    Answer:

$3x + 4y = 12$ using $mx+b$

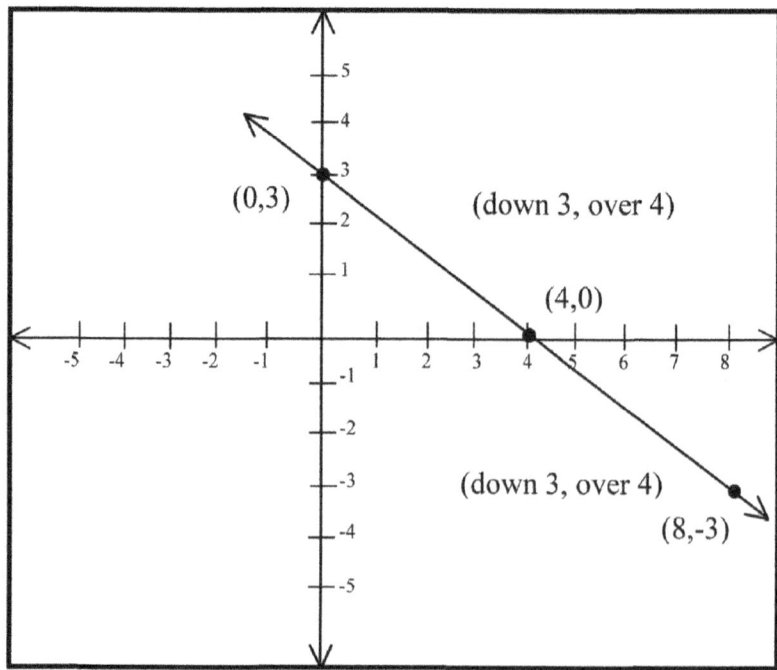

It is necessary to rewrite the equation in the form $y=mx+b$. First, subtract $3x$ from both sides of the equation, leaving $4y=-3x+12$. Notice how it is written: $-3x+12$, not $12-3x$. Although either of those expressions is correct, it is preferable to write it as $-3x+12$ so it starts to look like $y=mx+b$. Now, divide each term by 4. This yields $y=-\frac{3}{4}x+3$. Therefore, the y-intercept is 3 and the slope is $-\frac{3}{4}$. Start at the y-intercept $(0,3)$ and place a point there. Now, go down three units and to the right four units and place a second point there. Continue the process a third time to secure a check point.

Example 6

A new charter school opens up. It begins with a first-grade class of twenty-five students. Each year, it plans to add another grade and enroll fifty students in the new grade. Assuming all students re-enroll each year, how many students will there be in the school when the students now in first grade are seniors in high school?

Answer: 575

To solve this, take the information given in the problem and identify what parts of the equation $y=mx+b$ are given. The slope represents the change in the number of students each year. In this problem, the slope equals 50 because 50 new students are enrolled every year. Letting the first year the school is open correspond to a value of 0, the y-intercept, or b equals 25 because the initial first grade class consisted of 25 students. Now, the school must enroll students for eleven more years for the current first grades to reach their senior year. Therefore, $x=11$. Substituting these values into $y=mx+b$ gives $y=50(11)+25=550+25=575$ students.

Practice Problems

1. What are the slope and y-intercept of the line $5x+3y=15$?
   (a) $m=5;\ b=3$
   (b) $m=-5,\ b=-3$
   (c) $m=-\frac{5}{3};\ b=5$
   (d) $m=\frac{5}{3};\ b=15$
2. What is the slope of the line $x=-3$?
   (a) 1
   (b) 0
   (c) undefined
   (d) -3
3. When they were first manufactured, seventy-inch large-screen televisions cost $15,000. As technology improved, the cost for the television decreased $2,000 each year. Given that the first seventy-inch large-screen television was manufactured five years ago, what does it cost today?
   (a) $25,000
   (b) $5,000
   (c) $10,000
   (d) $13,000

4. The current population of the city of Pumpkinville is 6,500. Each year, it has been observed that 400 new residents move into the city, but 50 current residents move out. Assuming this trend of 400 new residents moving in and 50 current residents moving out continues for 10 years, what will be the population of Pumpkinville then?
   (a) 6,850
   (b) 65,000
   (c) 14,000
   (d) 10,000

Detailed answer key for the above problems:

1. Answer: (c) $m=-\frac{5}{3}$; $b=5$
   Rewrite the equation so that it is in the form $y=mx+b$. First, subtract $5x$ from both sides. Remember to write the resulting equation as $3y=-5x+15$ so it starts to take the proper form. Next, divide all terms by 3. This yields $y=-\frac{5}{3}x+5$. Therefore, $m=\frac{-5}{3}$ and $b=5$.
2. Answer: (c) undefined

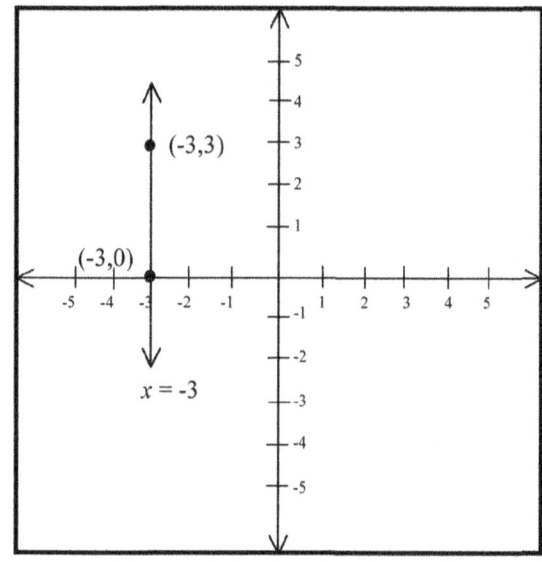

The line $x=-3$ is a vertical line. The slope of a vertical line (any line that takes the form $x =$ some number and does not contain a y) is always undefined. A way to remember this to associated the letters u (for undefined) and v (for vertical) by thinking of the UV rays given by the sun. Another way to remember, for those whose handwriting is sloppy, is to realize that u and v may look alike. To find the slope using the slope formula, $m=\frac{y_2-y_1}{x_2-x_1}$, let the point $(-3,0)=(x_1,y_1)$ and the point $(-3,3)=(x_2,y_2)$ Plugging these values into the formula $m=\frac{y_2-y_1}{x_2-x_1}$, yields $\frac{0-3}{3-3}$ or $\frac{-3}{0}$. Because it is impossible to divide by 0, the answer is undefined.
3. Answer: (b) $5,000
   To solve this, take the information given in the problem and identify what parts of the equation $y=mx+b$ are given. The slope represents the change in the price of the television each year. In this problem, the slope equals $-2,000$ because the price of

the television has decreased $2,000 each year. Letting the price five years ago correspond to a value of 0, the *y*-intercept, or *b* equals 15,000 because the price of the television five years ago was $15,000. The number of years of the price decreasing is given as 5. That is the value for *x*. The only value that is missing is *y*, the price of the television now. Plugging into the equation $y=mx+b$ results in $y=-2000(5)+15000=-10000+15000=5000$.

4. Answer: (d) 10,000

To solve this, take the information given in the problem and identify what parts of the equation $y=mx+b$ are given. The slope represents the change in the population of Pumpkinville each year. In this problem, this is a little tricky. There are four hundred new residents moving in (+400) but fifty current residents moving out (−50). The net change each year is 400−50 or +350 people each year. Thus, the slope equals 350. Letting the current population correspond to a value of 0, the *y*-intercept, or *b* equals 6,500. The number of years the population is changing is ten. That is the value for *x*. The only value that is missing is *y*, the population ten years from now. Plugging into the equation $y=mx+b$ results in $y=350(10)+6,500=3,500+6,500=10,000$.

GRAPHING PARALLEL AND PERPENDICULAR LINES

Two lines are parallel if the following conditions are met: (1) the lines have the same slope and (2) the lines have different *y*-intercepts.

Example 1

Graph the lines $y=2x+2$ and $y=2x-5$.
Answer:

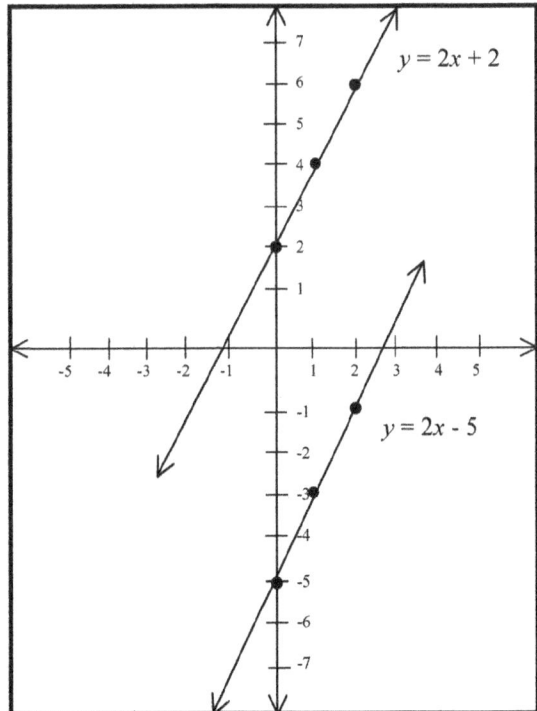

60   Chapter Four

Because the lines are already in the form $y=mx+b$, the easiest way to graph them is by inspection, or by using the y-intercept as the starting point and the slope to reach the next point. For the line $y=2x+2$, start at the point (0,2) and then go up two units and to the right one unit. Remember, if the slope is an integer (a whole number and their negative counterparts excluding 0, such as ... −3, −2, −1, 1, 2, 3), the denominator is always 1. Therefore, always go one unit to the right. For the line $y=2x-5$, start at the point (0,−5) and then go up two units and to the right one unit. The lines are parallel—they will never meet and the distance between them remains the same.

Two lines are perpendicular (make right angles) if the following condition is met: the slopes of the two lines multiply together to equal −1. The y-intercepts of the two lines may or may not equal each other. It does not matter. What does matter is that the slopes are negative reciprocals of one another. In other words, if one slope is +2, the other slope must be $-\frac{1}{2}$ because $+2(-\frac{1}{2})=-1$.

Example 2

Graph the lines $y=-4x$ and $y=\frac{1}{4}x+5$.
   Answer:

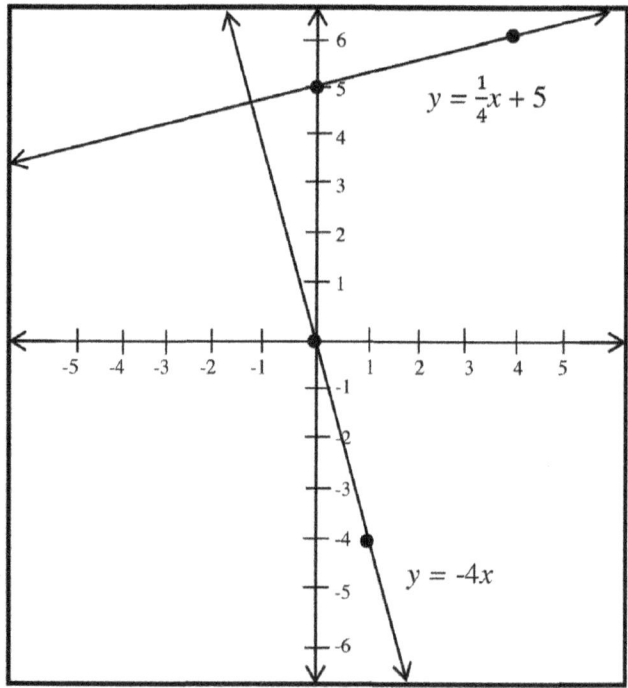

Because the lines are already in the form $y=mx+b$, the easiest way to graph them is by inspection, or by using the y-intercept as the starting point and the slope to reach the next point. For the line $y=-4x$, it may be helpful to write the equation as $y=-4x+0$. Therefore, the y-intercept equals 0, so start at the point (0,0) and then go down 4 units and to the right 1 unit. For the line $y=\frac{1}{4}x+5$, start at the point (0,5) and then go up one unit and to the right four units. The lines are perpendicular—their slopes multiply together to equal −1.

Practice Problems

1. Which of the following lines is parallel to the line $y=3x-7$?
   (a) $3x+y=7$
   (b) $\frac{1}{3}x+y=12$
   (c) $-3x+y=7$
   (d) $-\frac{1}{3}x+y=-7$
2. Which of the following lines is perpendicular to the line $y=-\frac{3}{4}x+17$?
   (a) $y=\frac{4}{3}x+17$
   (b) $y=-\frac{3}{4}x-17$
   (c) $y=-\frac{4}{3}x+17$
   (d) $y=-\frac{4}{3}x-17$
3. A given line passes through the points $(-5,12)$ and $(4,6)$. What is the slope of a line parallel to this line?
   (a) $-\frac{2}{3}$
   (b) $+\frac{2}{3}$
   (c) $+\frac{6}{9}$
   (d) it cannot be determined from the information given in the problem
4. A given line passes through the point $(2,4)$ and has a y-intercept of 6. What is the slope of a line perpendicular to this line?
   (a) -1
   (b) $+1$
   (c) undefined
   (d) 0

Detailed answer key for the above problems:

1. Answer: (c) $-3x+y=7$
   For lines to be parallel, they must have the same slope and different y-intercepts. The given line has a slope of $+3$, so the other line must also have a slope of $+3$. Put all the possible answers in the form $y=mx+b$ by adding the x term to both sides. Answer (a) becomes $y=-3x+7$; answer (b) becomes $y=-\frac{1}{3}x+12$; answer (c) becomes $y=3x+7$; and answer (d) becomes $y=\frac{1}{3}x-7$. The only answer with a slope of $+3$ is (c).
2. Answer: (a) $y=\frac{4}{3}x+17$
   For lines to be perpendicular, their slopes must multiply together to equal -1, or opposite reciprocals. In other words, the slopes must have different signs (one positive and one negative) and the numbers must be reciprocals or flip-flops of each other. Answers (c) and (d) can be quickly eliminated because they have negative slopes, and the slope of the given line is $-\frac{3}{4}$. The negative reciprocal of $-\frac{3}{4}$ is $\frac{4}{3}$; therefore the answer is correct.

3. Answer: (a) $-\frac{2}{3}$

First, find the slope of the line containing the points $(-5,12)$ and $(4,6)$. Use the slope formula $m=\frac{y_2-y_1}{x_2-x_1}$, where the point $(-5,12)$ equals (x_1,y_1) and the point $(4,6)$ equals (x_2,y_2). Plugging this into the formula gives $\frac{6-12}{4-(-5)}=\frac{-6}{9}$. This can be reduced to $-\frac{2}{3}$. For lines to parallel, they must have the same slope. So the parallel line will also have a slope of $-\frac{2}{3}$.

4. Answer: (b) +1

If a line has a y-intercept of 6, the point (0,6) lies on the line. Remember, the y-intercept is the value of y found by letting $x=0$. Now, find the slope of the line containing the points (2,4) and (0,6). Use the slope formula $m=\frac{y_2-y_1}{x_2-x_1}$, where the point (2,4) equals (x_1,y_1) and the point (0,6) equals (x_2,y_2). Plugging into the formula gives $\frac{6-4}{0-2}=\frac{-2}{2}=-1$. For lines to be perpendicular, their slopes must multiply together to equal -1. Therefore, the slope of the new line must be $+1$ because $(+1)(-1)=-1$.

SOLVING SYSTEMS OF LINEAR EQUATIONS

Suppose a problem asked for the solution to $x+6=10$. Clearly, the answer is $x=4$. Now, suppose a problem asked for the solution $x+y=10$. There are an infinite number of solutions. One person might say $x=5$ and $y=5$ because $5+5=10$, while another person might say $x=0$ and $y=10$ because $0+10=10$. What other solutions might be suggested? Think of one. Any two numbers that add to 10 will be correct.

Why does the first equation only have one solution and the second equation have an infinite number of solutions? The answer has to do with the number of variables in the problem. A linear equation with only one variable usually has only one solution. If two variables are involved in the problem, at least two equations are needed to determine a unique solution. These two equations are called a system of equations. To solve a system of equations, it is necessary to find values for both x and y.

There are three ways to solve systems of equations: graphically, by the addition/elimination method and by substitution.

Example 1

Solve the following system of equations graphically:

$$x+y=10$$
$$x-y=12$$

Answer:

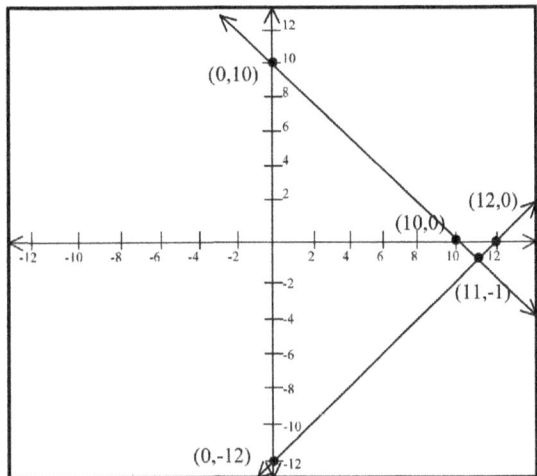

It is necessary to graph both lines on the same set of axes. To graph $x+y=10$, the fastest method is to use intercepts. If $x=0$, then $y=10$. Likewise, if $y=0$, then $x=10$. Graph the points (0,10) and (10,0) and draw the line connecting them. To graph $x-y=12$, rewrite the equation as $y=x-12$. (Subtract x from both sides, and then multiply each term by -1 so that $-y$ becomes $+y$. Rewrite so that the equation takes the form $y=mx+b$.) Start at the y-intercept, $(0,-12)$ and use the slope of $+1$ to go up one unit and to the right one unit to find the next point. Draw that line, and look for the point where the two lines intersect. This intersection point is the solution to the system of equations. This is the point: $(11,-1)$. Notice that both original equations are now true statements when $(11,-1)$ are substituted for x and y. For $x+y=10$, $11+(-1)=10$, and for $x-y=12$, $11-(-1)=12$. The final answer must satisfy both equations.

Example 2

Solve the following system using the addition/elimination method:

$$x+y=16$$
$$x-y=6$$

Answer: (11, 5)

To solve by the addition/elimination method, it is necessary to first put the equations in the form of $Ax+By=C$ (if they are not given in that form) and then write the second equation under the first one.

The equations are already in the proper form, so no rewrite is needed. In the addition/elimination method, the goal is to eliminate one of the variables by adding the two equations together. In this example, because y has a coefficient of $+1$ in the top equation and -1 in the bottom equation, when these two equations are added together, the y terms will cancel out. Add the two equations together to give the result of $2x=22$. This is now solvable because there is only one variable, x. Divide both sides by 2 to find that $x=11$. The problem is only complete when values for both x and y are known. Substitute the value of $x=11$ into either original equation to discover that $y=5$. If the two lines were graphed, as in the example above, the lines would intersect at the point (11,5).

Example 3

Solve the following system using the addition/elimination method:

$$3x+2y=16$$
$$5x+7y=34$$

Answer: (4,2)

This problem is a little harder to solve. Adding the two equations together as was done in the previous example will yield $8x+9y=50$. This is still one equation with two variables and cannot be solved. Decide which variable to eliminate. There is no wrong answer. It is possible to solve the problem by eliminating x or by eliminating y. To eliminate x, think of a number that both 3 and 5 go into. This is similar to finding a common denominator. Any number divisible by 3 and 5 will work, so choose 15. The goal is to make one of the coefficients of x positive 15 and the other coefficient of x negative 15 so that when the two equations are added together, the x terms will cancel out.

$$-5(3x + 2y = 16)$$
$$3(5x + 7y = 34)$$

$$-15x - 10y = -80$$
$$15x + 21y = 102$$

Adding the two equations together now means that the x terms cancel out, leaving $11y=22$ or $y=2$. Plugging the value of $y=2$ into any equation (the original equations of the problem or the new equations created by multiplying the top equation by -5 and the bottom equation by 3) results in $x=4$. The final answer is the ordered pair (4,2). Remember, it is impossible to be wrong when selecting which variable to eliminate. If the coefficients of a variable have different signs, think about selecting that variable because it will not be necessary to multiply one equation by a negative number. If the coefficients of a variable are multiples (such as a $2x$ term in the top equation and a $4x$ term in the bottom equation), think about selecting that variable because it will be necessary to only multiply one equation (in this case, the $2x$ term by -2, creating a $-4x$ term that will cancel with the $4x$ term). In other words, students can't pick wrong, but they can pick smart. It is left to the reader to re-solve the above problem by eliminating the y terms. (Hint: Think 14.)

Example 4

Solve the following system by the addition/elimination method:

$$x+y=16$$
$$x+y=6$$

Answer: There is no solution. The system is inconsistent.

Pick a variable to eliminate. To eliminate x, it is necessary for one coefficient to be positive 1 and the other to be negative 1. Now, pick an equation to change the coefficient of x to negative 1 by multiplying through by -1. Suppose the bottom equation is selected. The top equation remains $x+y=16$ but the bottom equation becomes $-x-y=-6$. Adding the two equations together results in both the x and y terms cancelling out, leaving 0 on the left-hand side of the equation and 10 on the right-hand side. $0=10$. Because this is a false statement, there is no solution. If the equations were rewritten in the form $y=mx+b$, the first equation would become $y=-x+16$ and the second equation would become $y=-x+6$. The equations have the same slopes (both $m=-1$) but different y-intercepts. Therefore, the lines are parallel and will never intersect. Hence, there is no solution.

Example 5

Solve the following system by the addition/elimination method:

$$x+\ y=16$$
$$2x+2y=32$$

Answer: There are an infinite number of solutions: all ordered pairs of the form $(x, 16-x)$

Pick a variable to eliminate. To eliminate x, it is necessary for one coefficient to be positive 2 and the other to be negative 2. Multiply the top equation by -2, yielding $-2x-2y=-32$. The bottom equation remains $2x+2y=32$. Adding the two equations together results in both the x and y terms and the constants cancelling out, leaving 0 on the left-hand side of the equation and 0 on the right-hand side. $0=0$. Because this is a true statement, there is an infinite number of solutions. All points on the line $x+y=16$ are solutions. If the equations were rewritten in the form $y=mx+b$, the first equation would become $y=-x+16$ and the second equation would become $2y=-2x+32$. Dividing by 2 gives $y=-x+16$. The equations have the same slopes (both $m=-1$) and the same y-intercepts (both $+16$). Therefore, the lines are coincident. In other words, when the first line is graphed, the second line will graph exactly on top of the first line. The two lines intersect at every point on the lines. Thus, there is an infinite number of solutions.

Example 6

Solve the following system by substitution:

$$x+y=16$$
$$y=2x+13$$

Answer: (1, 15)

The third method of solving systems of equations, substitution, works best when one variable is already given in terms of the other. In the problem above, the second equation gives y in terms of x because $y=2x+13$. To use substitution, go back to the first equation and wherever there is a y, substitute or replace the y with $2x+13$. Thus, the first equation becomes $x+(2x+13)=16$. This equation is now solvable because there is only one unknown, x. Combining like terms gives $3x+13=16$. Subtract 13 from both sides so that $3x=3$, and divide by 3, giving $x=1$. To find the value for y, plug $x=1$ into either original equation and determine that $y=15$. All systems can be solved by substitution. Sometimes, however, this can be a very messy way to solve the problem. In order to use substitution, it is necessary to write one of the equations so that either the x term is by itself on one side, and everything else is on the other side, or the y term is by itself on one side and everything else is on the other side. Often, this can result in a lot of fractions. For example, if $3x+7y=25$, to solve for x, subtract $7y$ from both sides and divide by 3. Therefore, $x=\frac{-7y+25}{3}$. Because this is not easy to work with, substitution is best left for those problems where one variable is already given in terms of the other.

Practice Problems

1. Solve the following system graphically:
   $2x+3y=6$
   $y=x+7$
   (a) $(3,-4)$
   (b) $(4,-3)$
   (c) $(-3, 4)$
   (d) $(-3, -4)$
2. Solve the following system by the addition/elimination method:
   $5x+3y=8$
   $4x+6y=10$
   (a) $(1,1)$
   (b) no solution
   (c) $(-1, -1)$
   (d) infinite number of solutions

3. Solve the following system by substitution:
   $4x-3y=-12$
   $x=-3y+12$
   (a) (4,0)
   (b) (0,4)
   (c) (−4,0)
   (d) (0, −4)
4. Solve by the method that is most appropriate for the problem. Four times the first number increased by 12 is equal to a second number. Two times the first number when added to three times the second number gives a result of −6. Write the answer as an ordered pair as follows: (first number, second number).
   (a) no solution
   (b) infinite number of solutions
   (c) (−3,0)
   (d) (0,−3)

Detailed answer key for the above problems:

1. Answer: (c) (−3, 4)

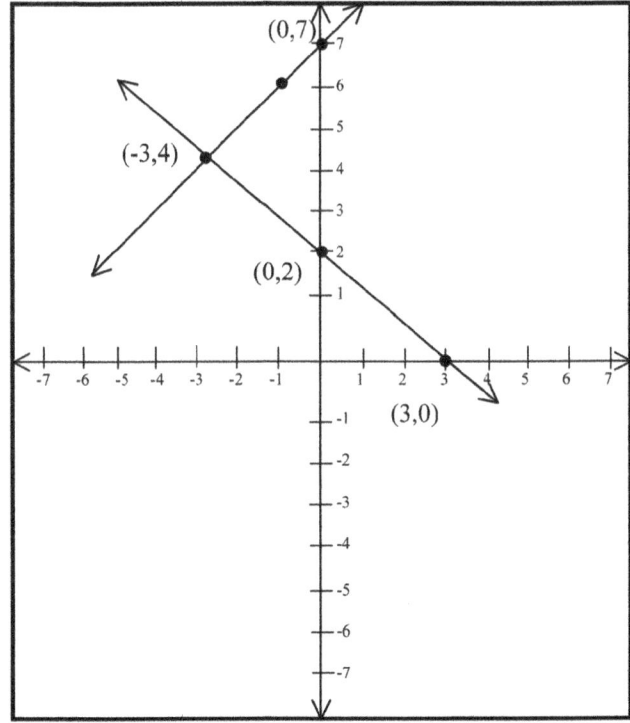

When graphing $2x+3y=6$, it is easiest to use intercepts. When $x=0$, $3y=6$ or $y=2$. Thus the point (0,2) is on the line. Likewise, when $y=0$, $2x=6$, or $x=3$ resulting in the point (3,0) being on the line. Graph the two points and draw the line that connects them. To graph $y=x+7$, graph by inspection because the equation is already in the form $y=mx+b$. Therefore, the y-intercept $=7$ (the point 0,7) and the slope $=1$, so go up one unit and to the right one unit to land on the next point. Draw that line, and notice how the lines intersect at the point $(-3,4)$.

2. Answer: (a) (1,1)

Remember, either x or y can be selected to eliminate. There is no right or wrong answer concerning which variable to pick. However, in this problem, it is faster to eliminate y because the coefficient of the second equation (6) is a multiple of the first equation (3). Whenever multiples are involved, it is necessary to only multiply through one equation. In this case, multiplying the top equation by -2 gives $-10x-6y=-16$. Now, the y terms will cancel when the new equation and the original second equation are added together. The result is $-6x=-6$, and dividing both sides by -6 gives a value of 1 for x. Plug this value for x into any of the three equations (the original two from the problem or the new one generated by multiplying the top equation by -2) to find that $y=1$.

3. Answer: (b) (0,4)

To solve by substitution, it is necessary to have one variable in terms of the other. In the problem, x is already given in terms of y. Therefore, simply substitute the value given for x in the other equation. Given the equation $4x-3y=-12$, replace the x with $-3y+12$. Now the equation reads $4(-3y+12)-3y=-12$. Distribute the 4 to yield $-12y+48-3y=-12$. Combine like terms on the left-hand side so that $-15y+48=-12$. Subtract 48 from both sides so $-15y=-60$. Dividing both sides by -15 gives a value for y of 4. Plug this into either equation to find that $x=0$.

4. Answer: (c) $(-3,0)$

This problem requires translating English sentences into algebraic equations and then solving a system. The first step is to identify the variables. Let $x =$ the first number and let $y =$ the second number. Consider the following sentence: four times the first number increased by 12 is equal to a second number. Translated to an equation, it would read $4x$ (four times the first number) + (increased by) 12 = (is equal to) y (the second number) or $4x+12=y$. Now, do the same for the second sentence: two times the first number when added to three times the second number gives a result of -6. The equation will be $2x+3y=-6$. Break the sentence into manageable chunks and write out each chunk separately. Go slowly and don't feel overwhelmed.

The best method to solve this system is substitution because one variable, y, is given in terms of the other in the first equation. Go to the second equation and replace y with $4x+12$. Therefore, $2x+3(4x+12)=-6$. Distribute to get $2x+12x+36=-6$. Combine like terms so that $14x+36=-6$. Subtract 36 from both sides: $14x=-42$. Divide both sides by 14, remembering that a negative divided by a positive is a negative, so that $x=-3$. Plug in -3 for x in either equation to find that $y=0$.

SOLVING AND GRAPHING LINEAR INEQUALITIES AND SYSTEMS OF LINEAR INEQUALITIES

An inequality is a mathematical statement that contains one of the four inequality signs. These signs include < (the sign for less than); > (the sign for greater than); ≤ (the sign for less than or equal to) and ≥ (the sign for greater than or equal to). The sign always points to the smaller number or opens on the bigger number. One good way to remember this is to draw teeth into the sign and remember that a hungry animal will always eat the most it can so make the sign open to the bigger number.

There are always two ways to write an inequality. For example, $5 < 7$ means the same as $7 > 5$. Think about this in terms of age. If Katie is twenty-four and Michael is twenty-six, it is equally correct to say that Katie is younger than Michael or Michael is older than Katie. When solving an inequality problem, don't get flustered if the answer is given in the other format.

The rules for solving an inequality are identical to the rules for solving an equation with one exception. Whenever a negative number is multiplied or divided to both sides of an inequality, the direction of the sign must be reversed. Clearly, $10,000 > 1$. However, if -1 is multiplied to both sides of this inequality, $-10,000 < -1$. It is perfectly acceptable to add or subtract the same quantity to both sides of an inequality. It is also fine to multiply or divide both sides of an inequality by any positive number. But, if a negative number is multiplied or divided to both sides of an inequality, don't forget to flip the sign.

Example 1

Solve: $5x > -15$

Answer: $x > -3$

A common mistake that is often made is to flip the sign whenever a negative number is involved. In the above example, to get the x by itself, it is necessary to divide both sides of the inequality by 5. But since the division is by a positive number, there is no need to flip the sign.

Example 2

Solve: $-25x < -100$

Answer: $x > 4$

Dividing both sides of the inequality by -25 results in x on the left-hand side and $+4$ on the right-hand side. Because the division was by a negative number, it is necessary to flip the direction of the inequality sign.

To graph a linear inequality, such as $3x+4y<12$, use the following procedure (shown on the next page). First, replace the inequality sign with an equality sign. Then, graph the line as above, using a dotted or broken line for $>$ or $<$ and a solid line for $\geq$ or $\leq$. Next, pick any point *not* on the line. When possible, use the point (0,0) since this is the easiest point to substitute. Plugging (0,0) into the original inequality, determine if the statement is true or false. If the statement is true, shade that point and all the points on that side of the line. If the statement is false, shade the other side of the line, which does not include that point. By using (0,0), it can be seen that $3(0)+4(0)=0$ is less than 12, so shade the region containing the point (0,0) or the region to the left of the line.

70   Chapter Four

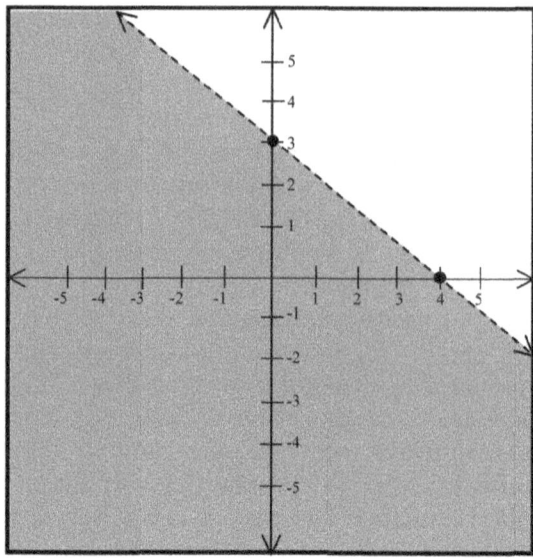

To graph a system of inequalities, graph each inequality separately using the above procedure. The portion of the graph that satisfies both inequalities will be the correct answer.

Example 3:

Graph $4x+2y<6$ and $y<4x+3$ on the same set of axes and shade the region of the graph which satisfies both inequalities.
  Answer:

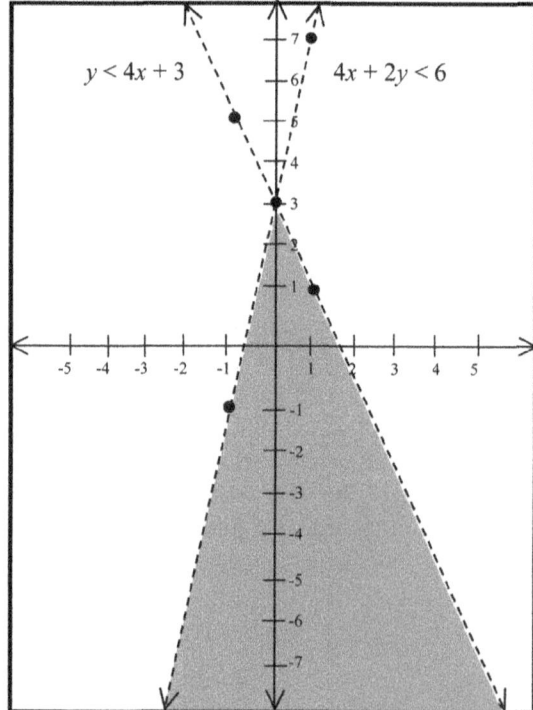

Algebra

First, graph the inequality $4x+2y<6$. Replace the inequality symbol with an equal sign so that $4x+2y=6$. The easiest way to graph this line is to use intercepts. When $x=0$, $y=3$, and when $y=0$, $x=\frac{3}{2}$. Plot these two points—(0, 3) and ($\frac{3}{2}$, 0)—and use a dotted line to connect them. Pick a point not on the line to determine where to shade. Using the point (0,0) and substituting it into the inequality results in a true statement because $0>-6$. Therefore, shade the region that contains the point (0,0).

Now, graph the inequality $y<4x+3$. The easiest way to graph this is by inspection. Begin at the y-intercept, (0,3) and go up four units and to the right 1 unit to determine the next point on the line. Connect the points using a dotted line. To determine which region to shade, again choose the point (0,0) and determine that 0 is, in fact, less than $4(0)+3$. Therefore, shade the region which contains the point (0,0).

The final answer is where the two regions overlap.

Practice Problems

1. Solve: $\frac{1}{2}x + \frac{3}{4} < \frac{2}{3}x + \frac{7}{8}$
   (a) $-\frac{3}{4}>x$
   (b) $-\frac{3}{4}\leq x$
   (c) $-\frac{3}{4}<x$
   (d) $-\frac{3}{4}=x$
2. Solve: $-\frac{2}{3}x \geq -12$
   (a) $x\geq 8$
   (b) $x\leq 8$
   (c) $x\geq 18$
   (d) $x\leq 18$
3. Graph: $5y\geq 15x+25$
4. Graph the following system: $4x+2y\geq 8$ *and* $5x-3y\leq 15$.

Detailed answer key for the above problems:

1. Answer: (c) $-\frac{3}{4}<x$
   To make the problem easier to handle, start by clearing the fractions. Find a common denominator for all the given denominators in the problem: 2, 4, 3, and 8. The least common denominator is 24. Next, multiply each term of the problem by 24. This results in a new (but equal) inequality of $12x+18<16x+21$. Subtract $12x$ from both sides, leaving $18<4x+21$. Finally, subtract 21 from both sides, giving $-3<4x$. The last step is to divide both sides by 4. Therefore, $-\frac{3}{4}<x$. Make sure to examine each answer carefully before marking one. It is tempting to select the first answer containing $-\frac{3}{4}$. But (a) and (b) are incorrect because they do not contain the proper inequality sign. And (d) is incorrect because it contains an equal sign. An inequality can never result in answer with an equal sign. An inequality must always have a final answer containing an inequality sign.
2. Answer: (d) $x\leq 18$
   To solve this inequality, it is necessary to move the coefficient of $-\frac{2}{3}$ from the left-hand side to the right-hand side. The easiest way is to multiply both sides of the inequality by the reciprocal of $-\frac{2}{3}$, which equals $-\frac{3}{2}$. Remember, a number and its reciprocal always share the same sign. The left-hand side of the inequality now

becomes +1*x* or *x* and the right-hand side becomes −12 times −$\frac{3}{2}$, which equals $\frac{36}{2}$=18. Because both sides of the inequality were multiplied by a negative number, it is mandatory to flip the direction of the inequality sign.
3. Answer:

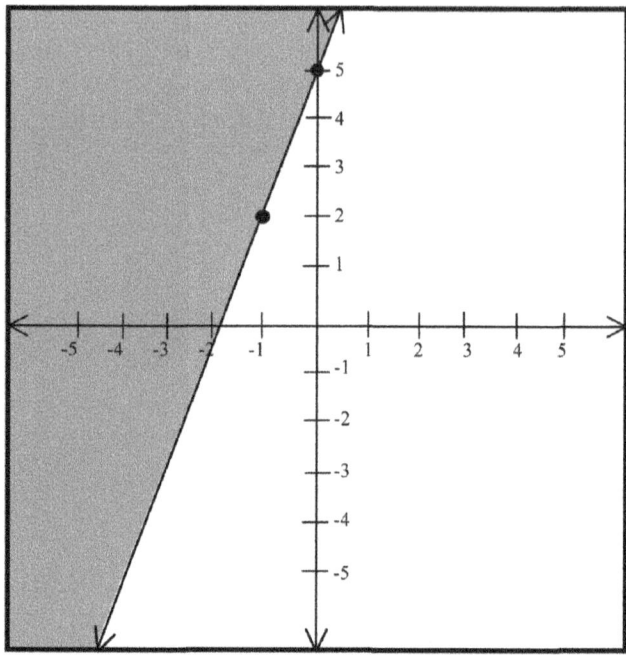

Remember the steps for graphing an inequality. First, change in the inequality sign to an equal sign. Therefore, $5y=15x+25$. The equation is almost in the form of $y=mx+b$ except that the coefficient of *y* is 5, not 1. Divide every term of the equation by 5 to result in $y=3x+5$. Don't forget to divide the 25 by 5. A common mistake that students make is to just divide the term with the *x* and not divide the constant. Whatever is done to one side of an equation must be done to the other side to maintain its equality. If the left-hand side is divided by 5, the entire right-hand side must also be divided by 5. Now graph the line using the *y*-intercept of (0,5) and the slope of 3. Go up three units and to the right one unit to get the next point on the line. Use a solid line because the original inequality used the greater than or equal to symbol. To determine which side of the line to shade, pick any point not on the line, using (0,0) when possible as is the case here. Plugging in (0,0) to the original inequality results in the false statement of 0≥25 so shade the region that does not contain the point (0,0).

4. Answer:

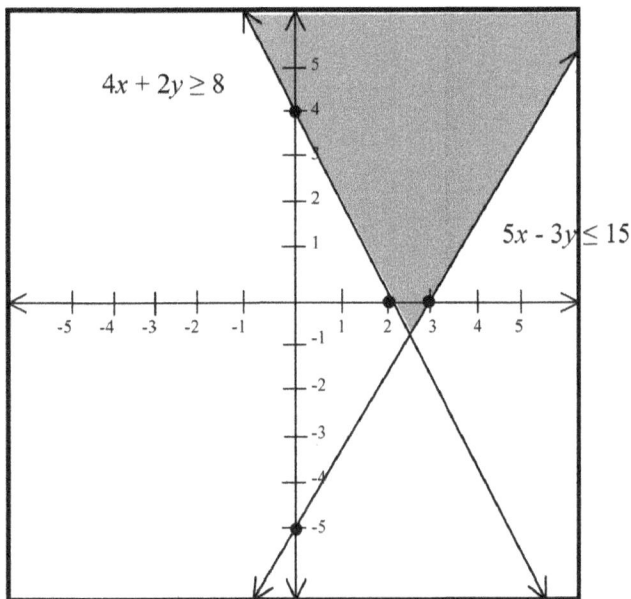

To graph these inequalities, replace the inequality symbols with equal signs. The fastest way to graph is to use intercepts. For the first inequality, plot the points (0,4) and (2,0), and connect them with a solid line. Pick the point (0,0) and plug into the original inequality to get the false statement that $0 \geq 8$. Therefore, shade the side of the line that does not contain the point (0,0). Likewise, for the second inequality, plot the points (0,−5) and (3,0) and connect them with a solid line. Again, pick the point (0,0) and plug it into the original inequality to get the true statement that $0 \leq 15$. This time, shade the side of the line containing the point (0,0). The final answer is the region where the two shadings overlap.

NONLINEAR FUNCTIONS AND REAL-WORLD SITUATIONS

One commonly seen nonlinear function on the test is the parabola. The equation of a parabola takes either the form $y=ax^2+b$ or $x=ay^2+b$. Note the squared term in both of these equations. Parabolas are u-shaped and may open up, down, or sideways.

Example 1

Graph $y=x^2$.

Answer:

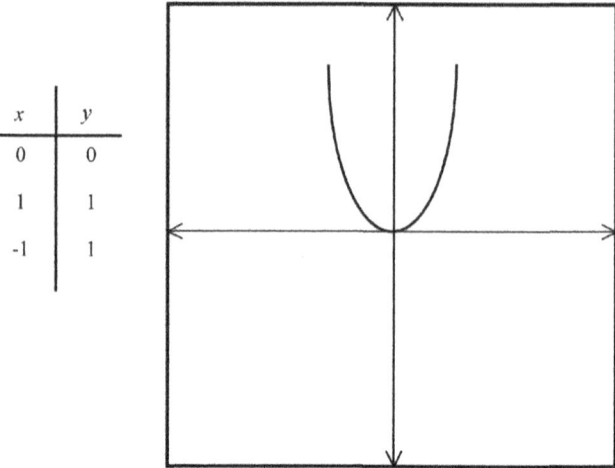

x	y
0	0
1	1
-1	1

Make a table of values for x, and find the corresponding values for y. When $x=-1$, $y=(-1)^2=1$; when $x=0$, $y=(0)^2=0$, and when $x=1$, $y=(1)^2=1$. Because of the squared term, the graph will be u-shaped, so connect the points in the shape of a u. Notice, when the coefficient of the x^2 term is positive, the parabola will open upward.

Example 2

Graph $y=-2x^2+3$.
Answer:

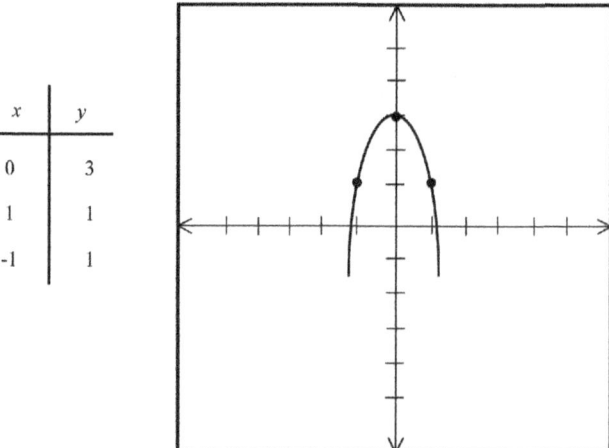

x	y
0	3
1	1
-1	1

Make a table of values for x, and find the corresponding values for y. When $x=-1$, $y=-2(-1)^2+3=-2+3=1$; when $x=0$, $y=-2(0)^2+3=0+3=3$, and when $x=1$, $y=-2(1)^2+3=-2+3=1$. Because of the squared term, the graph will be u-shaped, so connect the points in the shape of a u. Notice, when the coefficient of the x^2 term is negative, the parabola will open downward.

Example 3

Graph $x=y^2$.
Answer:

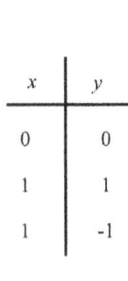
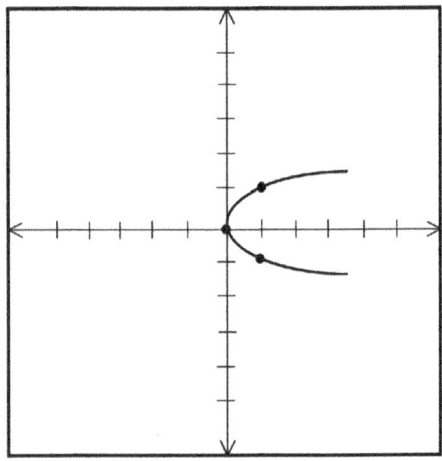

Make a table of values for y, and find the corresponding values for x. When $y=0$, $x=0^2=0$; when $y=1$, $x=1^2=1$, and when $y=-1$, $x=(-1)^2=1$. Because of the squared term, the graph will be u-shaped, so connect the points in the shape of a u. Notice, when the parabola is of the form $x=y^2$, it will open on its side.

Example 4

A cab charges a flat rate of $5.00 to pick up a passenger, $2.50 per mile for the first two miles, $1.50 per mile for the next three miles, and $1.00 per mile for any additional mileage. How much will a person pay to ride to the airport, which is fifteen miles from his current location, if he tips the driver 10 percent of the total cost?
Answer: $26.95
This problem contains a lot of information that needs to be broken into manageable bits. The rider will pay $5.00 just to be picked up. He is travelling a total of fifteen miles, so he will pay $2.50 for the first two miles. $2.50 per mile times 2 miles = $5.00. He still has 13 (15−2) miles to travel. For the next three miles, he will pay $1.50 per mile. $1.50 per mile times 3 miles = $4.50. But, he still has ten miles (13−3) to go. For those ten miles, he will pay $1.00 per mile, or $10.00. His cost for the trip, excluding the tip, is $5.00+$5.00+$4.50+$10.00=$24.50. A 10 percent tip on $24.50 is $2.45, so add $2.45 to the cost of $24.50 to get his final cost of $26.95.

Example 5

A certain bacteria called Dare-goo doubles in size every day. April, a mad scientist, goes into her lab on Monday, May 5th, and places 2 ounces of Dare-goo in a container. How many ounces of Dare-goo will April have on Friday, May 9th?
Answer: 32 ounces

Chart the amount of Dare-goo each day. On Monday, May 5th, April has 2 ounces. The next day, Tuesday, May 6th, the Dare-goo has doubled to 4 ounces. It doubles again on Wednesday, May 7th, to 8 ounces. On Thursday, May 8th, it doubles again to 16 ounces, and by Friday, May 9th, it has doubled to 32 ounces.

Practice Problems

1. Our mad scientist, April, has a second bacteria, Dickens-glob, which evaporates each day, leaving half the amount of the previous day. On Friday, May 9th, April discovers she has 4 pounds of Dickens-glob left in her container. How much Dickens-glob did she have on Monday, May 5th?
    (a) there is not enough information in the problem to determine how much she had on Monday.
    (b) $\frac{1}{4}$ pound
    (c) 20 pounds
    (d) 64 pounds
2. Michael loves stuffed animals. Every time his mother buys him one, his father puts $5.00 into Michael's college account, his grandmother buys him another stuffed animal and puts $10.00 into his college account, and his godmother buys him two stuffed animals but does not put money into his college account. One week, Michael's mother decides to spoil Michael and buys him one stuffed animal every day of the week. His father, grandmother, and godmother hear about this and do as described above. How many new animals (excluding the ones Mom bought for him) does Michael get and how much is deposited into his college account?
    (a) 21 animals and $75
    (b) 15 animals and $105
    (c) 21 animals and $105
    (d) 105 animals and $21
3. Marie is an extremely organized woman who follows her routine exactly every day. For every fifteen minutes of TV she watches, she reads for thirty minutes. For every thirty minutes she reads, she rides her exercise bike for ten minutes. If Marie watches two hours of TV one day, how long does she ride her exercise bike?
    (a) 1 hour
    (b) 1 hour 20 minutes
    (c) 1 hour 15 minutes
    (d) 2 hours
4. Graph: $y=3x^2+2$.

Detailed answer key for the above problems:

1. Answer: (d) 64 pounds
   In this problem, again chart out the information day by day. April had 4 pounds of Dickens-glob on Friday. This is half the amount she had on Thursday, so on Thursday, she had 8 pounds. This is half the amount she had on Wednesday, so on Wednesday, she had 16 pounds. This is half the amount she had on Tuesday, so on Tuesday, she had 32 pounds. This is half the amount she had on Monday, so on Monday, she had 64 pounds.

2. Answer: (c) 21 animals and $105

Again, this problem contains a lot of information. It is a good idea to read the problem several times. First, read it to get a sense of what the problem is about: stuffed animals and a college fund. Next, read it to start to get a sense of the details: his mom, dad, grandmother, and godmother all do something when his Mom buys Michael a stuffed animal. Finally, read it and figure out what the question is asking: If his mom buys a new stuffed animal every day of an entire week, what actions do his dad, grandmother, and godmother take?

Break the problem into two parts: how many new animals will Michael get from his grandmother and godmother, and how much money will his dad and grandmother deposit into his college account? Tackle the number of new animals first. His mom purchased a new stuffed animal for Michael every day for an entire week, which means that his mom bought him seven stuffed animals. Therefore, his grandmother will also buy him seven stuffed animals and his godmother will buy him fourteen stuffed animals since she buys two for every one his Mom buys for him. Therefore, Michael gets twenty-one more stuffed animals. Now, turn to the money deposited in his college account. His dad deposits $5.00 for every animal his mom buys, so his dad deposits $5.00 times 7, or $35.00. His grandmother deposits $10.00 for every animal his mom buys, so his grandmother deposits $10.00 times 7, or $70.00. Adding $35.00 and $70.00 means that Michael gets $105 deposited in his college account.

3. Answer: (b) 1 hour 20 minutes

The secret to getting word problems correct is to go slowly, break the problem into tiny bits, and work with each little bit of information. Start with what is given. Marie watched two hours of TV one day. This is the same as 120 minutes. For every fifteen minutes she watches TV, she reads for thirty minutes. How many fifteen-minute intervals did she watch TV? Divide 120 by 15 to find that she watched TV for eight fifteen-minute intervals. Now she must read for eight thirty-minute intervals. But, if she reads for eight thirty-minute intervals, she must ride her exercise bicycle for eight ten-minute intervals, or eighty minutes. Don't panic if eighty minutes does not appear as an answer. Change it into hours and minutes by dividing eighty minutes by sixty minutes per hour. This results in an answer of one hour with twenty minutes left over.

4. Answer:

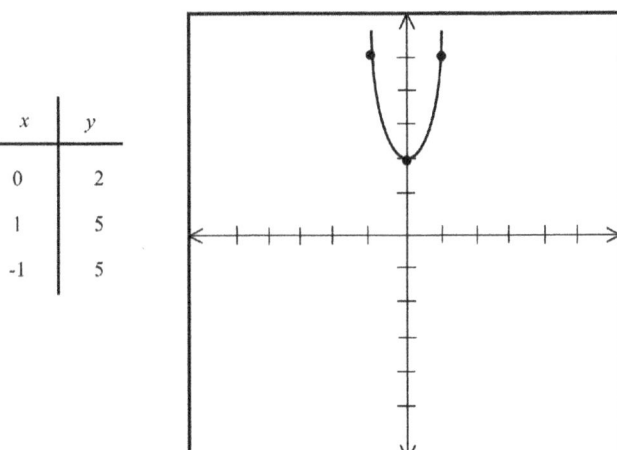

To graph this equation, pick values for x and solve for the corresponding values of y. When $x=0$, $y=3(0)^2+2=2$; when $x=1$, $y=3(1)^2+2=3+2=5$, and when $x=-1$, $y=3(-1)^2+2=3(1)+2=5$. Always pick values that are easy to plug into the original equation. Stay away from fractions and decimals—go with integers and/or whole numbers. Now graph the points (0, 2), (1, 5), and (−1, 5). Because of the squared term, recognize that the equation is a parabola, so connect the points in a u shape. Because the coefficient of the squared term is positive, the parabola will open upward.

Chapter Five

Measurement and Geometry

MEASUREMENT: THE U.S. CUSTOMARY SYSTEM

Length
12 inches (in.) = 1 foot (ft.)
36 inches = 1 yard
3 feet = 1 yard
5,280 feet = 1 mile
1,760 yards = 1 mile

Weight
16 ounces (oz.) = 1 pound (lb.)
2000 pounds = 1 ton (T)

Time
60 seconds (sec.) = 1 minute (min.)
60 minutes = 1 hour (hr.)
24 hours = 1 day
365 or 366 days = 1 year (366 if leap year, which is divisible by 4)

Capacity
4 pints (pt.) = 1 quart (qt.)
2 quarts = 1 gallon

Units of measure for the U.S. Customary System of weights and measures are based on the foot for length, the pound for weight, the second for time, and the fluid ounce for capacity. When converting any unit of measure from a smaller unit to a larger unit, divide the per-unit rate; when converting from a larger unit of measure to a smaller unit, multiply by the per-unit rate.

Example 1

Jason, Pete, and Andy measure 68 inches, 72 inches, and 70 inches respectively. What is their most precise average height in feet?

(a) 70 ft.
(b) 5.83 ft.
(c) 6 ft.
(d) there is not enough information given in the problem to determine the answer

Answer: (b) 5.83 feet

First, estimate each person's height in feet. Both Jason and Andy are a little under six feet and Pete is exactly six feet. Using reasonableness, answer choice (a) can be eliminated because the average of their heights cannot be seventy feet and answer choice (d) can be eliminated because there is definitely enough information available to solve this problem. Next, find their average height in inches. To find the average, add up all the three heights and divide by 3: $\frac{68+72+70}{3}$ inches. Now convert the inches to feet by dividing by 12 since there are twelve inches in one foot.

Example 2

A newborn weighs 9 lb. 2 oz. How many ounces is that?

(a) 144 oz.
(b) 146 oz.
(c) 108 oz.
(d) 110 oz.

Answer: (b) 146 oz.

Convert the nine pounds to ounces by multiplying by 16 oz./lb. This yields a result of 144, which is answer (a). Remember to add the two ounces to this answer to arrive at the correct answer of 146.

Practice Test Questions on U.S. Customary Measurement

1. One wall in Katie's bedroom measures ten feet. She wants to decorate the length of the wall with border paper. The border paper comes in rolls of one yard. How many rolls should she purchase so she has enough to complete the job?
   (a) 1
   (b) 2
   (c) 3
   (d) 4
2. Three baseball players weigh 180 lbs, 200 lbs, and 190 lbs respectively. What is their average weight in ounces?
   (a) 9,120
   (b) 11,875
   (c) 3,040
   (d) 1,187.5
3. A submarine descends one thousand feet in the Atlantic Ocean. This depth is approximately how many miles?
   (a) less than $\frac{1}{4}$ mile
   (b) between $\frac{1}{4}$ mile and $\frac{1}{2}$ mile
   (c) between $\frac{1}{2}$ mile and $\frac{3}{4}$ mile
   (d) more than $\frac{3}{4}$ mile

4. One box of books weighs fifty pounds. How many tons do four hundred boxes of books weigh?
   (a) less than ten tons
   (b) exactly ten tons
   (c) more than ten tons
   (d) exactly one hundred tons

Detailed answer key for the above problems:

1. Answer: (d) 4 rolls
   Since there are three feet in one yard, three rolls would cover nine feet of the wall. However, to cover ten feet, an additional roll is needed. Therefore, Katie would need to purchase four rolls to have enough.
2. Answer: (c) 3,040
   First, find the average weight of the three players by adding their weights and dividing by three. Their average weight is 190 lb. To change pounds to ounces, multiply by 16 oz./lb. This provides the correct answer of 3,040.
3. Answer: (a) Less than $\frac{1}{4}$ mile
   One mile is equal to 5,280 feet. First, estimate by multiplying one thousand by four to discover that it is not even a mile yet, so the only option that really makes sense is answer choice (a). Calculate $\frac{1}{4}$ mile by dividing 5,280 by 4, which is equal to 1,320 feet. Therefore, since 1,000 feet is less than 1,320 feet, the submarine must have descended less than $\frac{1}{4}$ mile.
4. Answer: (b) exactly ten tons
   If one box of books weighs fifty pounds, four hundred boxes of books weigh 400 times 50, or 20,000 pounds. To convert pounds to tons (smaller to larger), divide by 2,000 lbs./ton.

MEASUREMENT: THE METRIC SYSTEM

While the United States uses the U.S. Customary System of measure, many other countries have adopted the metric system. This system is based on powers of ten. The base unit of measure for length in the metric system is the meter. The base unit for capacity is the liter, and the base unit for weight is the gram.

By applying the metric system prefixes to the meter, the following units of measure are obtained.

The Metric System

KILO-	HECTO-	DEKA-	Base Unit—1	DECI-	CENTI-	MILLI-
King	Henry	Died	Meter—Length Liter—Capacity Gram—Weight	Drinking	Chocolate	Milk
× 1000	× 100	× 10		× 0.1	× .01	× .001

It is easy to convert among the various measures within the metric system. Since the units are based on power of ten, converting one metric measure of length to another involves moving the decimal point. To convert a larger unit to a smaller unit, the operation is multiplication, or moving the decimal point to the right. To convert a smaller unit to a larger unit, the operation is division, or moving the decimal point to the left.

Just as many students remember the phrase "Please Excuse My Dear Aunt Sally" or "PEMDAS" for the correct order of operations, students often use the following sentence to help remember how to convert metric units: King Henry Died (meter, liter, gram) Drinking Chocolate Milk.

King (kilo-) **H**enry (hecto-) **D**ied (deka-) (meter, liter, gram) **D**rinking (deci-) **C**hocolate (centi-)

Milk (milli-)

Example 1

Convert 17.68 kilometers to centimeters.
   Answer: 1,768,000 centimeters

Because kilometer is a larger unit than centimeter, the operation is multiplication. But when multiplying by a power of ten, move the decimal point to the right. Using the table above, start at kilometers and count how many places to centimeters. Since there are five places, move the decimal point to the right five places. Sometimes, as in this case, it is necessary to add extra zeroes to the unit being converted.

Write 17.68 kilometers as 17.6800000 kilometers (adding zeroes at the end of a decimal does not affect its value). Now move the decimal point five places to the right to arrive at the correct answer of 1,768,000 centimeters.

Example 2

Convert 800 grams to kilograms.
   Answer: 0.8 kilograms

Because gram is smaller than kilogram, and three places separate them, move the decimal point three places to the left: 800 grams = 0.8 kilograms.

Practice Problems

1. Three members of the men's basketball team are 1.8 meters, 2.0 meters, and 2.2 meters tall respectively. What is their average height in centimeters?
   (a) 20
   (b) 200
   (c) 2,000
   (d) 20,000
2. In July 1969, when American astronauts first walked on the moon, they had traveled a distance of approximately 384,000 kilometers. How many decimeters did they travel?
   (a) 3,840,000
   (b) 384,000,000
   (c) 3,840,000,000
   (d) 38,400,000,000

3. Cathy's new car weighs 2,500 kilograms. How many hectograms does Cathy's car weigh?
   (a) 25,000
   (b) 250
   (c) 250,000
   (d) 25
4. Arrange from smallest to largest: 9 grams, 85 dekagrams, 6020 milligrams, 54 decigrams.
   (a) 9 grams, 85 dekagrams, 6020 milligrams, 54 decigrams
   (b) 54 decigrams, 9 grams, 85 dekagrams, 6020 milligrams
   (c) 54 decigrams, 6020 milligrams, 9 grams, 85 dekagrams
   (d) 54 decigrams, 85 dekagrams, 9 grams, 6020 milligrams

Detailed answer key for the above problems:

1. Answer: (b) 200
   First, find the average height of the basketball players in meters: $\frac{1.8+2+2.2}{3}$. Now, convert two meters to centimeters by moving the decimal point two places to the right because meters is a larger unit than centimeters: 2.00 meters = 200 centimeters.
2. Answer: (c) 3,840,000,000
   As usual, the first decision to be made in solving the problem is whether the conversion is from a smaller unit to a larger unit or from a larger unit to a smaller unit. Because the conversion is from a larger unit (kilometers) to a smaller unit (decimeters), the operation is multiplication or moving the decimal point the appropriate number of places to the right. Kilometers and decimeters are four places apart, so the decimal point is moved four places to the right.
3. Answer: (a) 25,000
   Kilogram is one place further to the left than hectogram. To convert from kilogram to hectogram, the decimal point must be moved one place to the right.
4. Answer: (c) 54 decigrams, 6020 milligrams, 9 grams, 85 dekagrams
   In order to arrange these quantities from smallest to largest, the first step is to convert them to the same unit: 85 dekagrams = 850 grams, 6020 milligrams = 6.020 grams, and 54 decigrams = 5.4 grams. Now that all the quantities are in grams, simply arrange them from smallest to largest.

PERIMETER, AREA, VOLUME, AND SURFACE AREA

The perimeter of a geometric figure is the distance around the outside of the figure. Perimeter is measured in the U.S. Customary System and the Metric System units previously discussed. Perimeter answers are in inches, feet, yards, and miles in the U.S. Customary System and in kilometers, hectometers, dekameters, meters, decimeters, centimeters, and millimeters in the metric system.

The area of a geometric figure is the space occupied by the figure. Area is two-dimensional and is measured in inches^2, feet^2, yards^2, miles^2, meters^2, centimeters^2, etc.

The volume of a geometric figure is how much a figure can hold. Volume is three-dimensional and is measured in inches³, feet³, yards³, miles³, meters³, centimeters³, etc.

Surface area is the area of the outer surface of a three dimensional figure. Because it is a measure of area, its units include inches², feet², yards², miles², meters², centimeters², etc.

Study tip: Always watch the units given in the problem. If the problem is asking for area and the dimensions are given in different units, in order to solve the problem all the units must be the same. Look at the given answers and convert the units to the units in the answer. If the answers contain more than one unit, pick a unit and convert all the measures to that unit, realizing it might be necessary to convert your final answer to a different unit.

CIRCLES

The circumference of a circle is the distance around it. Circles do not have perimeters—they have circumferences.

A radius is any line segment from the center of the circle to the circumference. There are an infinite number of radii in every circle.

A chord is a line segment that connects two points on the circumference. One special chord, the diameter, is the line segment that passes through the center of the circle. All diameters are chords, but not all chords are diameters. The diameter is always the longest chord of the circle and equals twice the radius.

All circles contain 360°.

The circumference of a circle is found by using either of the following formulas: $C=\pi d$ or $C=\pi 2r$. Pi is an irrational number that may be approximated as 3.14 or $\frac{22}{7}$, d is the diameter of the circle, and r is the radius of the circle. The final answer will be in inches, feet, yards, etc.

The area of a circle is found by using the following formula: $A=\pi r^2$, where r is the radius of the circle. Remember, if given the diameter of the circle, it is necessary to divide that length by two in order to get the radius before applying the formula. The final answer for area is always measured in squared units: inches², feet², yards², etc.

Students can easily remember these formulas by the phrase "Cherry pie's delicious; apple pies are too": $C=\pi d$ and $A=\pi r^2$.

Example 1

Find the circumference of a circle with a diameter of nineteen feet in terms of π.

Answer: 19πft

$C=\pi d$, so simply substitute the value given for the diameter into the formula, since the circumference is to be in terms of π

Measurement and Geometry

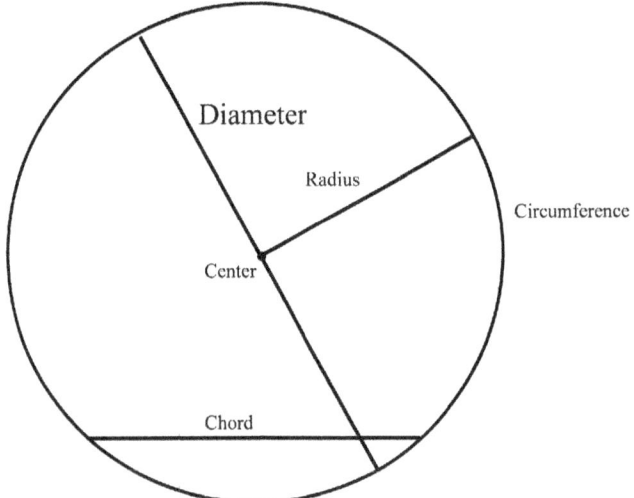

Example 2

Find the area of a circle with diameter 10 feet.
   Answer: 25π ft^2

Remember, to find area when given the diameter, it is first necessary to divide the diameter by two to find the radius. Here, the radius is ½ of 10 or 5. Now, use the formula $A=\pi r^2$, substituting the value of 5 in for r.

Practice Problems

1. The Malvern Gardening Club plants flowers in a circle with a radius of 18 inches. How many feet of fencing should the club buy to enclose their circular garden?
   (a) 36π ft
   (b) 18π ft
   (c) 3π ft
   (d) 9π ft
2. Katie bought her parents a circular clock for the kitchen of their new home. If the area of the clock is 16π in^2, what is the circumference of the clock?
   (a) 4π in
   (b) 8π in^2
   (c) 8π in
   (d) 4π in^2
3. Janet crocheted a circular rug with a circumference of 30π in. What is the area of the rug?
   (a) 15π in^2
   (b) 225π in^2
   (c) 950π in^2
   (d) 60π in^2

4. A circle has a circumference of 2π and an area of π. What is the circle's diameter?
   (a) 2
   (b) 4
   (c) 6
   (d) it cannot be determined from the information given in the problem

Detailed answer key for the above problems:

1. Answer: (c) 3π feet
   The problem gives the radius in inches but wants the circumference in feet, so the first step is to change eighteen inches into feet by dividing by 12 in/ft: $\frac{18}{12}=\frac{3}{2}$ ft. Now, plug that radius into the formula for circumference: $C=2\pi r=2(\frac{3}{2})$ft$\pi=3\pi$ ft.
2. Answer: (c) 8π inches
   Knowing that the area of the clock is 16π in^2 means that the radius of the clock must be four inches. The formula for circumference, when the radius is given, is $2\pi r$. Therefore, the circumference is $2\pi(4)$ or 8π in. Watch the units. It is easy to go too fast and mark answer (b) without seeing that answer (b) is in inches^2.
3. Answer: (b) 225π inches^2
   If the circumference is 30π inches, the radius must be 15 inches because $C=2\pi r$. Area is found by using the formula $A=\pi r^2$ so substitute the value of 15 in for r to arrive at the correct answer of 225π inches^2.
4. Answer: (a) 2
   At first glance, it might look like there is not enough information given to solve this problem. However, if the circumference is 2π, the radius must be 1 because $C=2\pi r$ and the only value for r that works is 1. Likewise, if the area is π, the radius must be 1 because $A=\pi r^2$ and the only value for r that works is 1. Thus, the diameter must be 2.

TRIANGLES

A triangle is a three-sided closed figure. The sum of the angles of any triangle is equal to 180°.

Example 1

What is the measure of the third angle of a triangle if one angle measures 40° and the other angle measures twice that?

Answer: 60°

One angle measures 40° and the angle that measures twice that must measure 80°. Together, the two angles total 120°. Another 60° is needed so that the angles total 180°.

Triangles are classified according to the number of equal sides and equal angles they have. A scalene triangle has no equal sides and no equal angles.

An isosceles triangle has two equal sides and two equal angles. The equal sides are opposite the equal angles and the equal angles are opposite the equal sides. If given an isosceles triangle, it is necessary to only know the measure of one angle to calculate the other two angles.

Example 2

The measure of one angle of an isosceles triangle is 100°. Find the measure of the other two equal angles.

Answer: 40°, 40°

Because one angle measures 100°, the remaining two angles must total 80° to satisfy the requirement that a triangle contains 180°. If two equal angles total 80°, each angle must be $\frac{1}{2}$ of that, or 40°.

An equilateral triangle has three equal sides. An equiangular triangle has three equal angles. All equilateral triangles are equiangular, and all equiangular triangles are equilateral. A triangle cannot be just equiangular or just equilateral. Once the triangle has one of those properties, it automatically has the other property.

Each angle of an equiangular/equilateral triangle measures 60° because when 180° is divided into three equal parts, each part must be 60°. It does not matter how long the sides of the equilateral/equiangular triangle are—each angle always measures 60°.

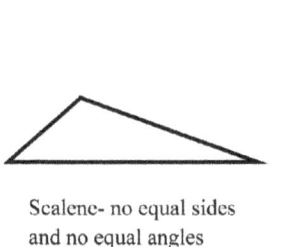

Scalene- no equal sides and no equal angles

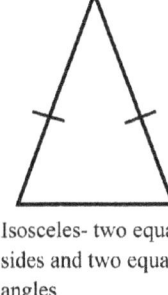

Isosceles- two equal sides and two equal angles

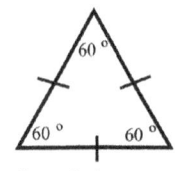

Equilateral- three equal sides

Equiangular- three equal angles

All angles equal 60°

A right triangle contains one right angle. The two sides that make up the right angle are called the legs and the side opposite the right angle is called the hypotenuse. The hypotenuse is always the longest side of a right triangle.

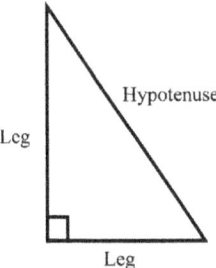

The Pythagorean Theorem states a relationship among the sides of the right triangle. According to the Pythagorean Theorem, hypotenuse^2 = leg^2 + leg^2. It does not matter in what order the legs are added because addition is commutative—the order of the addition does not effect the final answer. Remember, the answer found by using the above formula will result in the measurement of that side being squared—it is necessary to take the square root of that answer to find the actual measurement of the side. If the hypotenuse is given, to find a missing leg, use the formula hypotenuse² − leg² = leg².

Example 1

Find the hypotenuse of a right triangle whose legs measure 6 inches and 8 inches.
   Answer: 10 inches
   Use the formula hypotenuse² = leg² + leg² and substitute 6 and 8 for the legs. Therefore, hypotenuse² = 6² + 8² = 36+ 64 = 100. The hypotenuse does not measure one hundred inches—the hypotenuse² measures 100 inches. Take the square root of both sides (what number times itself equals one hundred?) to find the final answer of 10 inches.

Example 2

Find the leg of a right triangle whose hypotenuse is 17 feet and whose other leg is 15 feet.
   Answer: 8 feet
   To find the leg of the triangle, substitute the values that are given and solve for what is unknown.

$$a^2 + b^2 = c^2$$

$$15^2 + b^2 = 17^2$$

$$225 + b^2 = 289$$

$$289 - 225 = 64$$

$$b^2 = 64$$

$$b = 8$$

   Take the square root of both sides to find that the leg equals eight feet.

Two special right triangles are handy to know for the test. The 45-45-90 triangle is also known as the isosceles right triangle because two angles are equal, resulting in two legs being equal.

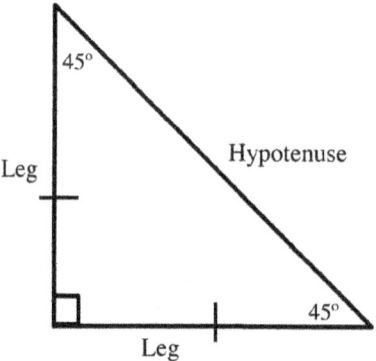

In a 45-45-90 right triangle, let the legs measure x and the hypotenuse of that triangle measure $x\sqrt{2}$. This can be shown by applying the Pythagorean Theorem: hypotenuse$^2 = x^2 + x^2 = 2x^2$. Taking the square root of both sides leads to the conclusion that the hypotenuse $= x\sqrt{2}$.

Example 3

Find the hypotenuse of an isosceles right triangle whose legs each measure seven.
   Answer: $7\sqrt{2}$.
Memorize the fact that in an isosceles right triangle (45-45-90), the hypotenuse always equals the leg$\sqrt{2}$.

The other special right triangle is the 30-60-90 right triangle.

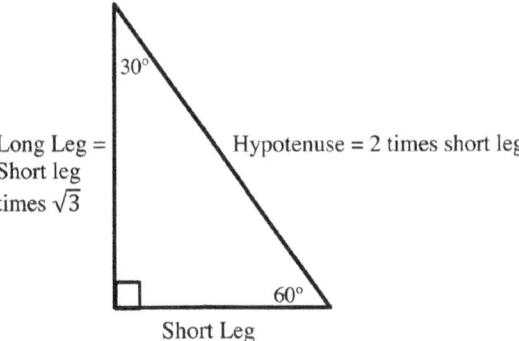

In a 30-60-90 right triangle, make sure to first label the sides. As expected, the side opposite the right angle is the hypotenuse, or longest side. It is necessary to distinguish the legs because they are not equal since they are not opposite equal angles. It is customary to call the leg opposite the 30-degree angle the short leg, and the side opposite the 60-degree angle the long leg. In every 30-60-90 right triangle, the following relationships are true:

The hypotenuse equals two times the length of the short leg and the long leg equals the short leg times $\sqrt{3}$. To remember which special triangle uses $\sqrt{2}$ and which one uses 3, keep in mind that the one with two equal legs uses the $\sqrt{2}$, and the one whose angles are all divisible by 3 uses the $\sqrt{3}$.

Example 4

Find the perimeter of a 30-60-90 triangle with short leg = 8 inches.
   Answer: $24 + 8\sqrt{3}$ inches

To find the perimeter, it is first necessary to find the lengths of the other leg and hypotenuse. The long leg = short leg$\sqrt{3}$ = $8\sqrt{3}$, and the hypotenuse = 2(short leg) = 2(8) = 16. Therefore, the three sides of the triangle measure 8, 16, and $8\sqrt{3}$. To find the perimeter of a triangle, simply add the lengths of the three sides. P = length of side 1 + length of side 2 + length of side 3. Add these three sides, resulting in a final answer of $24 + 8\sqrt{3}$ inches.

To find the area of a triangle, use the formula $A=\frac{1}{2}bh$, where b is the measure of the base of the triangle and h is the measure of the height of the triangle. Some students may have learned the formula $A=\frac{1}{2}ba$, where a represents the altitude of the triangle. The height and altitude of a triangle are the same; it is always the measure straight up and down, or perpendicular from the base—never the measure of a side that slants.

According to the triangle inequality, the sum of any two sides of a triangle is always greater than the third side.

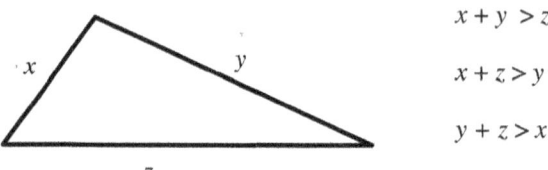

$$x+y>z$$
$$x+z>y$$
$$y+z>x$$

The sum of any two sides of a triangle is always greater than the third side

Two triangles are congruent if the three sides of one triangle equal the three sides of the other triangle and the three angles of one triangle equal the three angles of the other. If triangles are congruent, one triangle will fit exactly on top of the other triangle. They will be perfect matches.

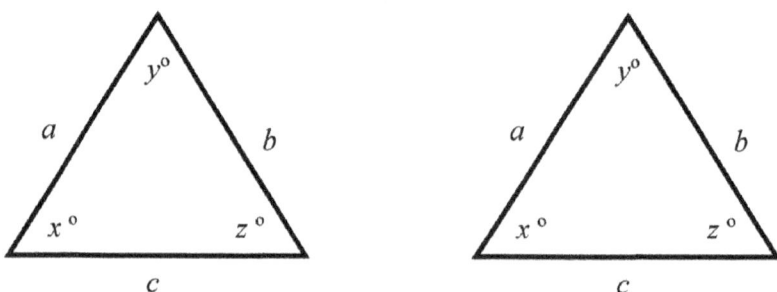

Two triangles are similar if the three angles of one triangle equal the three angles of the other triangle and the sides are in proportion. In other words, all three sides of one triangle are multiplied by the same positive number to create the three sides of a new triangle. For example, right triangles with sides 3, 4, and 5 and 6, 8, and 10 are similar. The angles are the same, and the sides of the second triangle are all two times the sides of the first triangle.

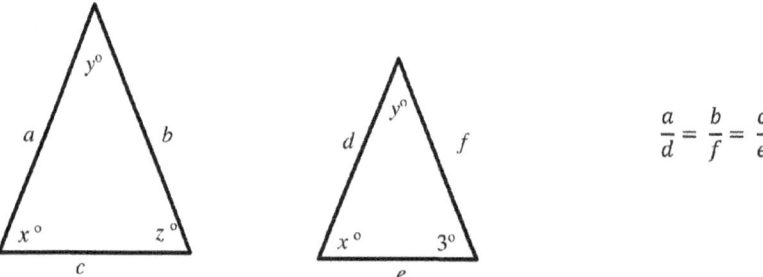

$$\frac{a}{d}=\frac{b}{f}=\frac{c}{e}$$

Practice Problems

1. According to the triangle inequality, which of the following cannot be the sides of a triangle?
    - I. 1, 1, 2
    - II. 1, 2, 3
    - III. 2, 9, 6
   - (a) I only
   - (b) I and II only
   - (c) III only
   - (d) I, II, and III
2. Find the area of a right triangle with one side of three feet, a height of four feet, and a hypotenuse of five feet.
   - (a) 12 ft^2
   - (b) 6 ft^2
   - (c) 25 ft^2
   - (d) 9 ft^2
3. A triangle has sides 6 in, $\frac{2}{3}$ ft and $\frac{1}{3}$ yd. Find its perimeter in inches.
   - (a) 7 in
   - (b) 26 in
   - (c) 24 in
   - (d) 18 in
4. Find the area of a triangle with base 8 in and height 1 ft.
   - (a) 4 in^2
   - (b) 4 ft^2
   - (c) 48 in^2
   - (d) 48 ft^2

Detailed answer key for the above problems:

1. Answer: (d) I, II, and III
   The first observation to be made about this problem is that it includes a thought-reverser. The question is asking for the set of the numbers that "cannot" be the sides of a triangle. Therefore, we are looking for three sides that do not satisfy the triangle inequality rule, which states that the sum of any two sides of a triangle is always greater than the third side. Option I does not satisfy the triangle inequality rule because 1 + 1 is *not* greater than 2, II does not satisfy the triangle inequality because 1 + 2 is *not* greater than 3, and III does not satisfy the triangle inequality because 2 + 6 is *not* greater than 8.
2. Answer: (b) 6 ft^2
   Notice that this problem contains information that is not needed to solve it. When finding the area of any triangle, the formula is $A=\frac{1}{2}bh$. The measure of the hypotenuse is not used in the formula. Therefore $A=\frac{1}{2}(3 \times 4)$ or $A=\frac{1}{2}(12)=6$ ft^2.
3. Answer: (b) 26 in
   The major issue in this problem concerns units. One side is given in inches, one side in feet, and the third side in yards. Because all the answers are given in inches,

covert all the sides to inches. $\frac{2}{3}$ ft times $12\frac{in}{ft}$=8 in, and $\frac{1}{3}$ yd times $36\frac{in}{yd}$ =12 in. The perimeter is the sum of the three sides = 6 in + 8 in + 12 in = 26 in.

4. Answer (c) 48 in²

The area of a triangle is found by the formula $\frac{1}{2}bh$. However, as in other problems, there is a units issue. The base is given in inches, while the height is given in feet. It is easier to convert one foot to twelve inches than to convert eight inches to a fractional part of a foot. Therefore, $A = \frac{1}{2}$ (8 in)(12in) = 48 in².

QUADRILATERALS

A quadrilateral is a four-sided closed figure whose angle sum is equal to 360°. There are several different kinds of quadrilaterals.

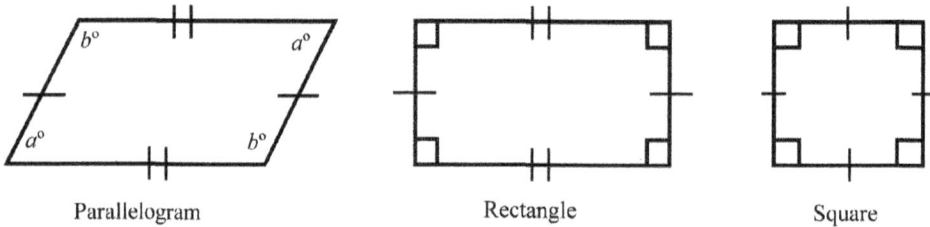

Parallelogram   Rectangle   Square

A parallelogram is a quadrilateral with the following properties:

1. Opposite sides are parallel. (They lie in the same plane and have no points in common.)
2. Opposite sides are equal in length.
3. Opposite angles are equal in measure.
4. Consecutive angles (angles that follow each other in the parallelogram, moving either clockwise or counterclockwise) always total 180°.

To find the perimeter of a parallelogram, simply add the four sides. To find the area of a parallelogram, use the formula $A = bh$ or ba, where b is the length of the base, h is the height, and a is the altitude. Remember, height and altitude mean the same thing and must always run perpendicular to the base.

A rectangle is a quadrilateral with the following properties:

1. Opposite sides are parallel.
2. Opposite sides are equal.
3. All angles equal 90°.

The formula for the perimeter of a rectangle may be expressed in one of two ways: $P=2l+2w$ or $P=2(l+w)$, where l is the length of the rectangle and w is the width of the rectangle. The area of a rectangle is given by the formula $A=lw$, where l is the length of the rectangle and w is the width of the rectangle.

Measurement and Geometry 93

A square is a quadrilateral with the following properties:

1. Opposite sides are parallel.
2. All sides are equal in length.
3. All angles equal 90°.

To find the perimeter of a square, use the formula $P=4s$, where s is the length of the side of the square. The area of a square is given by formula $A=s^2$, where s is the length of the side of the square.

Practice Problems

1. A garden is in the shape of a parallelogram. One side measures fifteen feet and the other side measures eighteen yards. Fencing costs $5.15 per foot. How much does it cost to fence in the entire garden?
   (a) less than $700
   (b) exactly $700
   (c) more than $700
   (d) more than $1,000
2. A rectangle with sides that are whole numbers has an area of 48 in². Which of the following is *not* a possible value for its perimeter?
   (a) 30 in
   (b) 52 in
   (c) 98 in
   (d) 28 in
3. What is the perimeter of a parallelogram that has the same dimensions as a rectangle whose sides are whole numbers and whose area is 19 ft²?
   (a) 19 ft
   (b) 40 ft²
   (c) 40 ft
   (d) the answer cannot be determined because not enough information is given
4. A rectangle has a perimeter of 46 ft. One side measures 2 in. What does the other side measure?
   (a) 23 in
   (b) 23 ft
   (c) 274 in
   (d) 276 ft

Detailed answer key for the above problems:

1. Answer: (c) More than $700
   Because the fencing is priced in feet, it is best to convert the eighteen yards to fifty-four feet. (18 yards times 3 feet/yard = 54 feet.) The perimeter is the sum of the four sides. $P=54+54+15+15=138$ feet. To find the total price of the fencing, multiply 138 feet by $5.15/ft = $710.70.

2. Answer: (a) 30 in

   For a rectangle to have an area of 48 in², its length and width must multiply to 48. The whole number pairs that multiply to 48 are: 1 and 48, 2 and 24, 3 and 16, 4 and 12, and 6 and 8. To find the perimeter of these possible rectangles, add the sides and multiply by 2. 1+48=49 × 2=98; 2+24=26 × 2=52; 3+16=19 × 2=38; 4+12=16 × 2=32; and 6+8=14 × 2=28. Hence, the answer is (a) 30 in since 30 does not appear in the list above.

3. Answer: (c) 40 ft

   For a rectangle to have an area of 19 sq. ft., it must measure 1 ft by 19 ft. 19 is a prime number, so there are no other numbers that when multiplied together will yield an answer of 19. Therefore, the parallelogram must be 1 ft wide and 19 ft long, so its perimeter is found by adding the four sides: 1 ft + 1 ft + 19 ft + 19 ft = 40 ft. Notice the units in answer (b). Make sure to select the proper units. Because the problem asks for perimeter, the answer must be in feet, not feet².

4. Answer (c) 274 in

   The perimeter is given in feet, while the measurement of the side is given in inches. First, change the forty-six feet to inches by multiplying 46 ft × 12 in/ft = 552 in. The formula for the perimeter of a rectangle is given by $P=2(l+w)$. Therefore, $552=2(l+2)$ or $552=2l+4$. Subtracting 4 from both sides yields $548=2l$. Dividing both sides by 2 results in $l = 274$ in.

IRREGULAR FIGURES/SHADED REGIONS

To find the perimeter or area of an irregular figure or shaded region, it is often helpful to break the figure into known shapes and then combine the familiar shapes accordingly.

Example 1

A circle with a radius of nine inches is inscribed in a square with a side of eighteen inches. Find the area of the shaded region in terms of π.

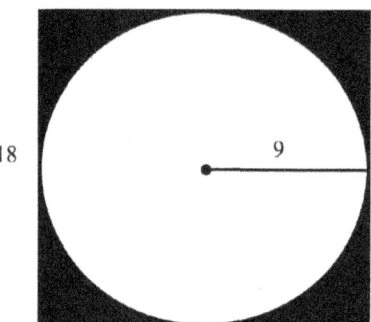

Answer: 324 in²−81π in²

To find the area of the shaded region, find the area of the square: $18^2=324$ in². Now, find the area of the circle $=\pi(9)^2=81\pi$ in². The area of the shaded region is the difference between these two areas: 324 in²−81π in².

Example 2

Joe builds a sign in the shape of an ice cream cone to advertise his new store. The base of the sign, the cone, is an equiangular triangle with sides that are twelve feet long. The scoop of ice cream at the top is a semicircle attached to the cone. Find the perimeter of the sign in terms of π.

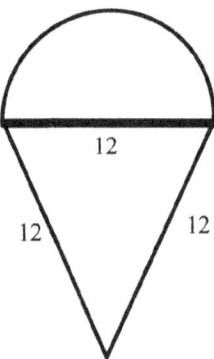

Answer: $24 + 6\pi$ ft

To find the perimeter, it is necessary to add the two sides of the cone (12 ft + 12 ft = 24 ft) to the circumference of the semicircle with a diameter of 12. The third side of the triangle also serves as the diameter of the circle. The circumference of the semicircle is $\frac{1}{2}\pi d = \frac{1}{2}\pi(12 \text{ ft}) = 6\pi$ ft. The total perimeter is the sum of the two parts: 24 ft + 6π ft.

Rectangular Prisms

A rectangular prism is simply a box. It has three dimensions: length, width, and height. To find the volume of a rectangular prism, multiply length times width times height. Remember, all units of measure must be the same before multiplication.

To find the surface area (SA) of a rectangular prism, use the formula $SA = 2lw + 2lh + 2wh$, where l, w, and h are the length, width, and height, respectively, of the prism.

Example 1

Find the volume in feet^3 of a rectangular prism with length 1 foot, width 6 inches, and height 1 yard.

Answer: $\frac{3}{2}$ ft³

Because the answer is to be in feet³, it is first necessary to change all measures to feet. 6 in = $\frac{1}{2}$ ft, and 1 yd = 3 ft. Therefore, the volume is 1 ft times $\frac{1}{2}$ ft × 3 ft or $\frac{3}{2}$ ft³.

Example 2

Find the surface area (*SA*) in ft² of a rectangular prism with length 2 yd, width 12 in, and height 2 ft.

Answer: 40 ft²

Again, if the answer is to be given in ft², change all units to ft. 2 yd = 6 ft and 12 in = 1 ft.

$SA = 2lw + 2lh + 2wh = 2(6)(1) + 2(6)(2) + 2(1)(2) = 12 + 24 + 4 = 40$ ft²

Practice Problems

1. A rectangular prism has volume of 12 ft³. Which of the following *cannot* be its dimensions?
   (a) 1 ft × 1 ft × 12 ft
   (b) 12 in × 144 in × 12 in
   (c) 5 ft × 2 ft × 2 ft
   (d) 3 ft × 4 ft × 1 ft
2. The length of a rectangular prism is twice its width. Its height is three times its width. If the volume is 48 ft³, what is its length?
   (a) 2 ft
   (b) 4 ft
   (c) 6 ft
   (d) the length cannot be determined from the information given
3. The length of a rectangular prism is twice its width. Its height is three times its width. If the surface area is 22 ft², what is its height?
   (a) 1 ft
   (b) 2 ft
   (c) 3 ft
   (d) the height cannot be determined from the information given
4. All of the following rectangular prisms have the same volume except:
   I. 3 ft × 2 ft × 1 ft
   II. 1 yd × 12 in × $\frac{2}{3}$ yd
   III. $\frac{1}{3}$ yd × 36 in × 24 in
   IV. 36 in × 24 in × 12 in
   (a) I
   (b) II
   (c) I and II
   (d) they all have the same volume

Detailed answer key for the above problems:

1. Answer: (c) 5 ft × 2 ft × 2 ft
   All the others have a volume of 12 ft³. Answer (c) has a volume of 20 ft³.

2. **Answer: (b) 4 ft**
Let w represent the width. The length is twice the width, or $2w$. The height is three times the width, or $3w$. The volume is found by multiplying length times width times height. Volume = 48 ft³=$(2w)(w)(3w)=6w^3$. Divide both sides by 6, leaving 8 ft³=w^3 or w = 2 ft. The length is 2 times 2 ft, or 4 ft.

3. **Answer: (c) 3 ft**
Let w represent the width. The length is twice the width, or $2w$. The height is three times the width, or $3w$. The surface area is found by the formula $SA=2(lw)+2(lh)+2(wh)=2(2w)(w)+2(2w)(3w)+2w(3w)=4w^2+12w^2+6w^2=22w^2=22$ ft². Therefore, $w^2=1$ or w = 1 and height is $3w$ or $3(1)$ = 3 ft.

4. **Answer: (d) they all have the same volume**
Convert all units to feet.
   I. Has a volume of 6 ft³.
   II. Also has a volume of 6 ft³ after converting 1 yd to 3 ft, 12 in to 1 ft, and $\frac{2}{3}$ yd to 2 ft.
   III. Can be converted to $\frac{1}{3}$ yd = 1 ft, 36 in = 3 ft, and 24 in = 2 ft. The volume is 6 ft³.
   IV. Can be changed to 36 = 3 ft, 24 in = 2 ft, and 12 in = 1 ft. The volume is 6 ft³.

Since all four answers are in essence 3 ft by 2 ft by 1 ft, all of them have the same volume.

SPHERE

A sphere is a ball-shaped figure. The volume of a sphere is found by formula $V=\frac{4}{3}\pi r^3$, where r is the radius of the sphere. Remember, π is the irrational number that may be approximated as $\frac{22}{7}$, or 3.14.

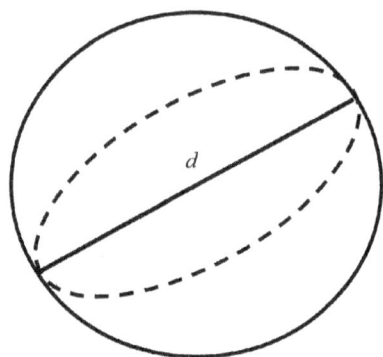

The surface area of a sphere is a given by the formula $(SA)=4\pi r^2$, where r is the radius.

Example 1

Find the volume of a sphere with diameter 2 ft.
Answer: $\frac{4}{3}\pi$ ft³
If the diameter of the sphere is 2 ft, its radius is 1 ft: $\frac{4}{3}\pi(1)^3=\frac{4}{3}\pi$ ft³.

Example 2

Find the surface area of a sphere with radius $\frac{1}{2}$ ft.
  Answer: π ft^2
  Surface area is found by $4\pi r^2 = 4\pi(\frac{1}{2})^2 = 4\pi\frac{1}{4} = \pi$ ft^2.

Practice Problems

1. A golf ball has a diameter of 0.6 in. What is its volume?
   (a) $\frac{6}{123}\pi$ in^3
   (b) $\frac{9}{125}\pi$ in^3
   (c) $\frac{12}{125}\pi$ in^3
   (d) $\frac{9}{250}\pi$ in^3
2. A ball has a volume of 36π ft^3. What is its diameter?
   (a) 3 ft
   (b) 6 ft
   (c) 9 ft
   (d) 1.5 ft
3. The surface area of a sphere is 144π in^2. What is its volume?
   (a) 288π in^3
   (b) 36π in^3
   (c) 144π in^3
   (d) 72π in^3
4. A sphere's surface area in square inches and its volume in cubic inches equal the same number. What is the radius of the sphere?
   (a) 9 in
   (b) 6 in
   (c) 3 in
   (d) it cannot be determined from the information given

Detailed answer key for the above problems:

1. Answer: (d) $\frac{9}{250}\pi$ in^3
   If the diameter is 0.6 inches, the radius is half of that, or 0.3 inches. Now apply the volume formula: $V = \frac{4}{3}\pi r^3$. Therefore, $V = \frac{4}{3}\pi(\frac{3}{10})^3 = \frac{4}{3}\pi\frac{27}{1000} = \frac{108}{3000}\pi$. Reduce to lowest terms by dividing numerator and denominator by 12: $V = \frac{9}{250}\pi$ in^3.
2. Answer: (b) 6 ft.
   Set up an equation using the volume formula, with 36π equaling the volume and solve for the radius: $36\pi = \frac{4}{3}\pi r^3$. Divide both sides by π, leaving $36 = \frac{4}{3}r^3$. Multiply both sides by $\frac{3}{4}$ to get the r^3 by itself. Therefore, $27 = r^3$ or $r = 3$. Don't be fooled into marking answer choice (a). The question is asking for the diameter, so double the radius to arrive at the final correct answer of 6 ft.
3. Answer: (a) 288π in3
   Set up an equation using the surface area formula, with 144π equaling the surface area and solve for the radius: $144\pi = 4\pi r^2$. Divide both sides by π, leaving $144 = 4r^2$.

quadrant IV. The only point that is not in any quadrant and has its own special name is the point (0,0). It is known as the origin.

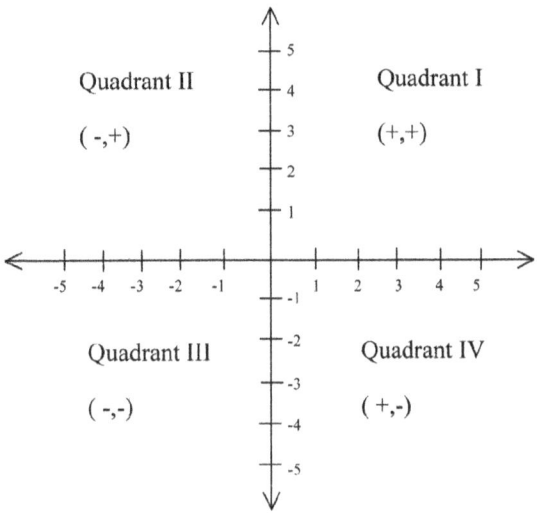

Given two points, it is possible to find the midpoint of the line connecting the two points, the distance between the two points, and the slope of the line containing the two points. Each value is found by using the appropriate formula.

Suppose the points (2,1) and (5,5) are given.

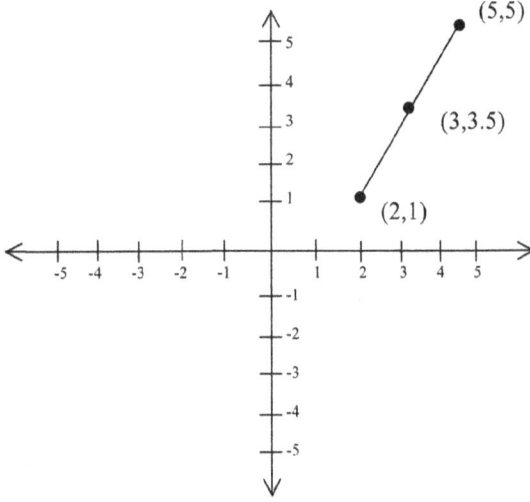

To find the midpoint of the line connecting the two points, plug the given values into the midpoint formula.

$$\left(\frac{x_1+x_2}{2}, \frac{y_1+y_2}{2}\right)$$

In the example, the midpoint equals $(\frac{7}{2}, \frac{6}{2})$ or $(3\frac{1}{2}, 3)$.

To find the distance of the line between the two points, plug the given values into the distance formula.

$$\sqrt{(x_2-x_1)^2+(y_2-y_1)^2}$$

The distance is the square root of the difference of the *x* values, squared, plus the difference of the *y* values, squared.

In the example, the distance is the square root of $(5-1)^2+(5-2)^2=$ the square root of $16 + 9$, or the square root of 25, which is 5.

TRANSFORMATIONS OF POINTS AND GEOMETRIC FIGURES

There are three types of transformations that may appear on the test. The first, called a reflection, is the result of flipping a figure. Think of a reflection as a mirror image.

Example 1

What are the coordinates of the point A(2,5) after it is reflected over the *x*-axis?
Answer: A (2,–5)

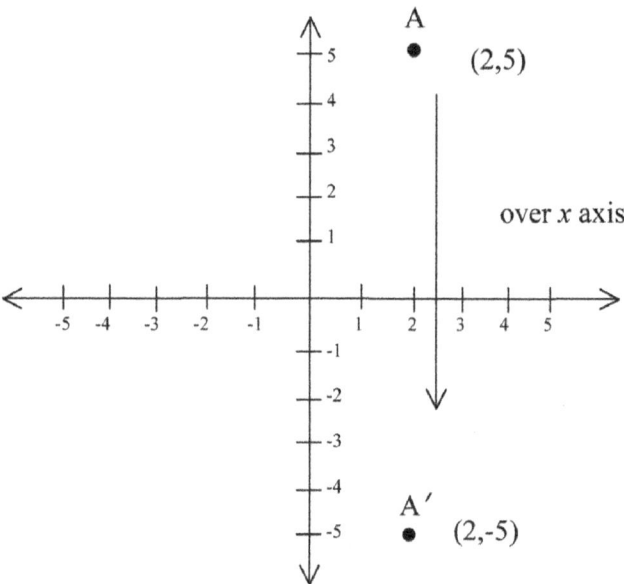

When a point is reflected over the *x*-axis, the *x* value in the ordered pair remains the same, but the value for *y* changes its sign. Similarly, when a point is reflected over the *y*-axis, the *y* value in the ordered pair remains the same, but the value for *x* changes its sign. Whatever axis is being reflected over, remember that the value in the ordered pair for that value remains the same.

Now, divide both sides by 4, yielding $36=r^2$. Therefore, the radius equals 6 in. Now, use the volume formula with r equal to 6 to find the final answer. $V=\frac{4}{3}\pi(6)^3=\frac{4}{3}\pi(216)=288\pi$ in^3. Make sure to cancel 3 into both the numerator and denominator to make calculations easier.

4. Answer: (c) 3 in

   Set up an equation in which the surface area equals the volume. Therefore, $4\pi r^2=\frac{4}{3}\pi r^3$. Divide both sides by πr^2, leaving $4=\frac{4}{3}r$. Now, multiply both sides of the equation by $\frac{3}{4}$, yielding a final answer of $3=r$.

ANGLES AND LINES

Angles are composed of two rays that meet a point called the vertex of the angle. They are usually designated by three letters, with the vertex letter always being the middle letter of the three. Sometimes, angles are just called by their vertex letter.

Angles are classified according to their size. Angles measuring less than ninety degrees are known as acute angles. An easy way to remember this is to visualize the angle as "a cute little angle." Angles measuring exactly ninety degrees are called right angles. Never assume that an angle measures ninety degrees. Nice mathematical things happen when an angle measures ninety degrees. These things don't happen if the angle is, for example, eighty-nine degrees or ninety-one degrees.

An obtuse angle measures more than 90 degrees but less than 180 degrees. It is a fat angle. The trick to remembering this is to think of the word *obese* and compare it to the word *obtuse*. The two words have many of the same letters in the same order. To be obese is to be fat, so an obtuse angle is a fat angle.

A straight angle measures exactly 180 degrees. Every straight line is a straight angle and therefore contains 180 degrees.

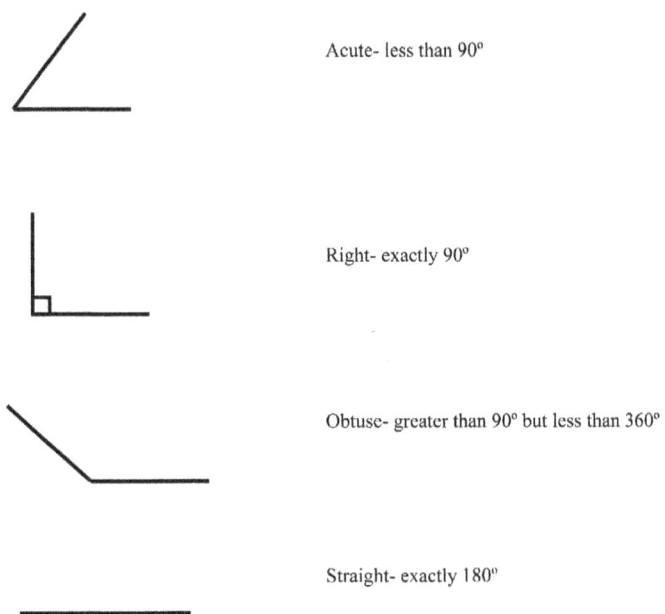

Acute- less than 90°

Right- exactly 90°

Obtuse- greater than 90° but less than 360°

Straight- exactly 180°

Two angles are said to be complementary if they add together to total ninety degrees. If one angle is given, to find its complement, simply subtract the known angle from ninety degrees. For example, to find the complement of a fifty-five-degree angle, subtract 55 from 90 and find that the complement is 35 degrees. The only angle that is equal to its complement is the forty-five degree angle.

Two angles are said to be supplementary if they add together to total 180 degrees. If one angle is given, to find its supplement, simply subtract the known angle from 180 degrees. For example, to find the supplement of a fifty-five degree angle, subtract 55 from 180, yielding a result of 125 degrees. The only angle that is equal to its supplement is the ninety-degree angle. Sometimes, students confuse the concepts of complementary and supplementary. One handy way to remember them is to think of the letter *s* standing for both *straight* and *supplementary*. Because a straight angle contains 180 degrees, supplementary angles must total 180 degrees.

Perpendicular lines intersect at right angles. The symbol for perpendicular lines is ⊥.

In figure 5.16, line 1 ⊥ line 2 because they meet at right angles. Each of the four angles formed by their intersection will measure 90 degrees,

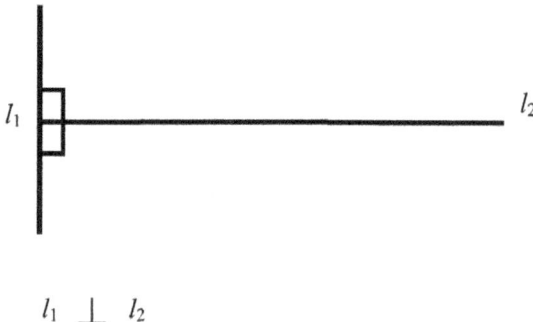

$l_1 \perp l_2$

Parallel lines never meet and the distance between them remains the same. The classic example of parallel lines is railroad tracks. Railroad tracks appear to go on forever and the distance between the tracks at any points on the tracks is the same. If two parallel lines are cut by a third line, called the transversal, eight angles are formed. These eight angles are related in a variety of ways.

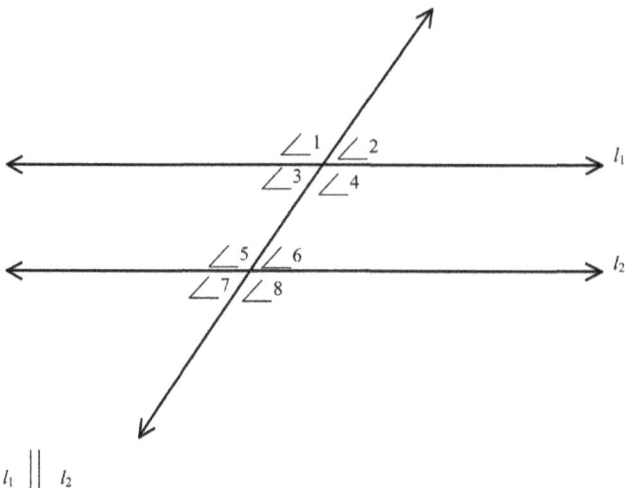

$l_1 \parallel l_2$

Angle 1 = angle 8, and angle 2 = angle 7. These are called alternate exterior angles. *Alternate* means they are on different (or alternating) sides of the transversal and *exterior* means they lie outside the parallel lines.

Angle 4 = angle 5, and angle 3 = angle 6. These are called alternate interior angles. Again, *alternate* means they are on different (or alternating) sides of the transversal and *interior* means they lie inside the parallel lines.

Angle 1 = angle 4, angle 2 = angle 3, angle 5 = angle 7, and angle 6 = angle 8. These are called opposite angles because each angle is opposite the other.

Angle 1 = angle 5, angle 2 = angle 6, angle 3 = angle 7, and angle 4 = angle 8. These are called corresponding angles because they hold the same relative position in the diagram. Angle 1 is to the left of the transversal and above one of the parallel lines. Angle 5 is to the left of the transversal and above the other parallel line.

Angle 3 + angle 5 = 180 degrees, and angle 4 + angle 6 = 180 degrees. Same-side interior angles are supplementary. Remember, *supplementary* means two angles that add to 180 degrees. *Interior* means the angles lie within the parallel lines, on the same side of the transversal.

For all of the above relationships to hold, the two lines must be parallel. If any one of the eight angles is given, the other seven angles can be determined given the one. Fill in the given angle, and the other three angles equal to that angle using the above definitions, and then find the supplement of the given angle and fill in the remaining four angles with that measurement.

Example 1

In the figure below, find the measures of the missing angles. Angle 2 = $x+37$. Angle 3 = $2x+13$.

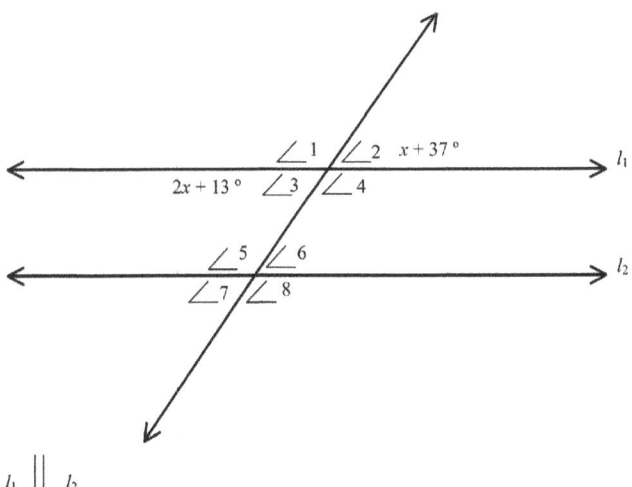

Answer: angles 2, 3, 6, and 7 = 61 degrees, and angles 1, 4, 5, and 8 = 119 degrees.

Angle 2 and angle 3 are opposite angles, so they are equal to each other. Therefore, $x+37=2x+13$. Subtracting x from both sides yields $37=x+13$. Subtract 13 from both sides to find that $x=24$. Therefore, angle 2 = 24+37, or 61 degrees, which implies that angles 3, 6, and 7 are all 61 degrees, and the remaining angles (angles 1, 4, 5, and 8) equal the supplement of 61 or 119 degrees.

COORDINATE GEOMETRY

Imagine a number line going from left to right (negative numbers on the left, 0, and positive numbers on the right) and a second number line crossing perpendicularly from the horizontal line at 0, and going up and down (negative numbers on the bottom and positive numbers on the top). This display is known as the Cartesian Coordinate System, named for the French mathematician Rene Descartes, who invented it. The number line that goes from left to right is called the x-axis, while the number line that goes from top to bottom is called the y-axis. It is easy to remember which axis is the y-axis. When you write the letter, notice how the tail of the y goes up and down—the y-axis also goes up and down.

Associated with the Cartesian coordinate system are points, also known as an ordered pair. An ordered pair takes the form (x,y) where x is where the point would cross the x-axis and y is where the point would cross the y-axis if lines were extended from the point. An ordered pair is "ordered" because it is in alphabetical order. Just as x comes before y in the alphabet, x comes before y in an ordered pair. Another way to remember this is to visualize an elevator. The elevator is the y-axis. Before one can go up and down the elevator (obtaining a value for y), one must enter the elevator (finding a location on the x-axis).

Example 1

Graph the ordered pair (2,5).
   Answer:

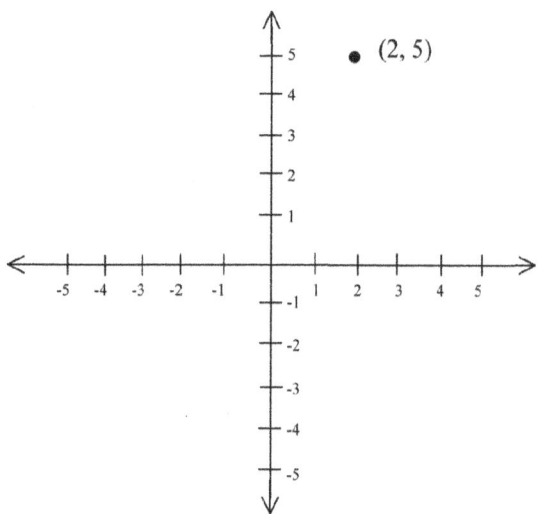

Always begin at the point of origin, which is (0,0). Now go to the x-axis to positive 2. Now, go up five units on the y-axis. That is the point (2,5).

The Cartesian Coordinate system is divided into four sections called quadrants. The quadrants are numbered by Roman numerals in a counterclockwise fashion. In quadrant I, both x and y take on positive values. In quadrant II, x is negative, while y is positive. Both values are negative in quadrant III. x is positive and y is negative in quadrant IV. The only point that is not in any quadrant and has its own special name is the point (0,0). It is known as the origin.

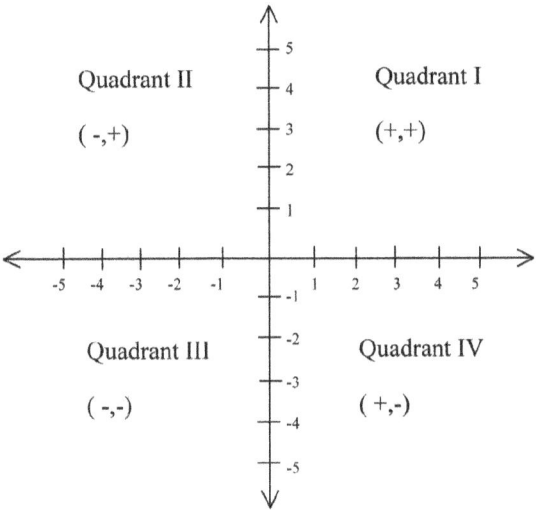

Given two points, it is possible to find the midpoint of the line connecting the two points, the distance between the two points, and the slope of the line containing the two points. Each value is found by using the appropriate formula.

Suppose the points (2,1) and (5,5) are given.

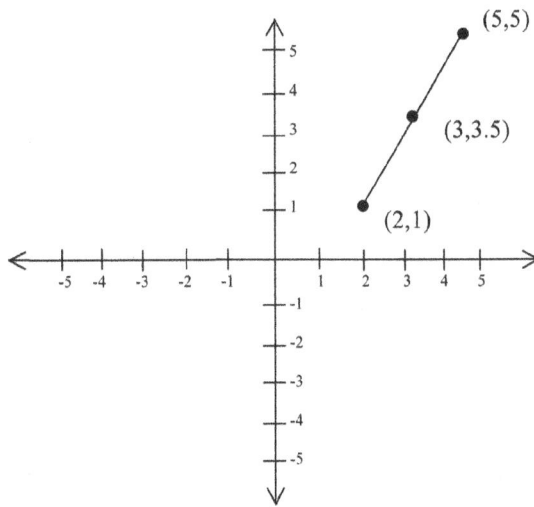

To find the midpoint of the line connecting the two points, plug the given values into the midpoint formula.

$$\left(\frac{x_1 + x_2}{2}, \frac{y_1 + y_2}{2}\right)$$

In the example, the midpoint equals $(\frac{7}{2}, \frac{6}{2})$ or $(3\frac{1}{2}, 3)$.

To find the distance of the line between the two points, plug the given values into the distance formula.

$$\sqrt{(x_2-x_1)^2+(y_2-y_1)^2}$$

The distance is the square root of the difference of the *x* values, squared, plus the difference of the *y* values, squared.

In the example, the distance is the square root of $(5-1)^2+(5-2)^2=$ the square root of $16 + 9$, or the square root of 25, which is 5.

TRANSFORMATIONS OF POINTS AND GEOMETRIC FIGURES

There are three types of transformations that may appear on the test. The first, called a reflection, is the result of flipping a figure. Think of a reflection as a mirror image.

Example 1

What are the coordinates of the point A(2,5) after it is reflected over the *x*-axis?
    Answer: A (2,–5)

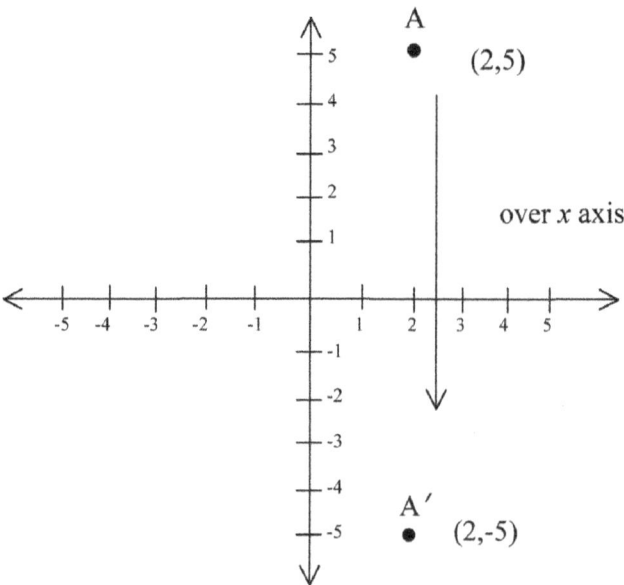

When a point is reflected over the *x*-axis, the *x* value in the ordered pair remains the same, but the value for *y* changes its sign. Similarly, when a point is reflected over the *y*-axis, the *y* value in the ordered pair remains the same, but the value for *x* changes its sign. Whatever axis is being reflected over, remember that the value in the ordered pair for that value remains the same.

Example 2

What are the coordinates of the point (3,4) after it is reflected over the line $y = x$?
  Answer: (4,3)

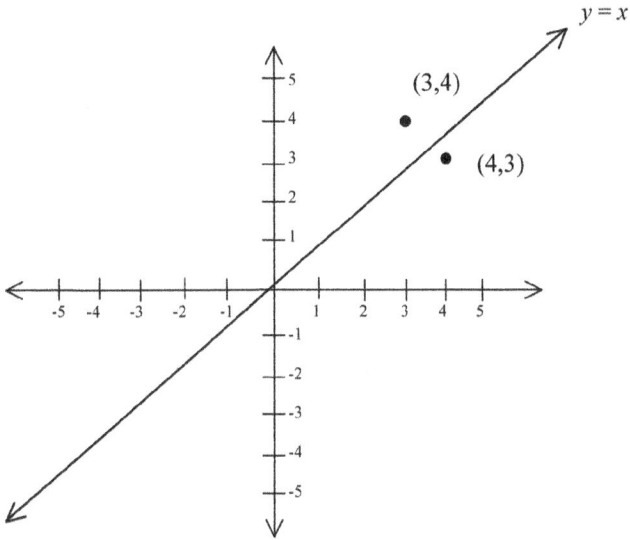

When reflecting over the line $y = x$ simply interchange the values of x and y. In other words, the new point is (y,x). Similarly, when reflecting over the line $y = -x$, the point (x,y) will reflect into the point $(-y,-x)$.

A translation, the second type of transformation, is a slide of the point or figure.

Example 3

Slide the point (–5,3) three units in the positive x direction and five units in the negative y direction. What would be the new point under this translation?
  Answer: (–2,–2)

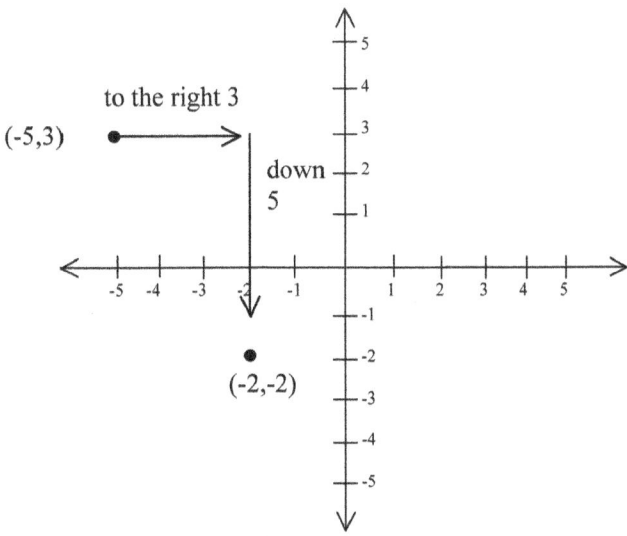

The original point (−5,3) now becomes (−5 + 3, 3 − 5), which equals (−2,−2).

The third type of transformation, known as a rotation, involves moving the figure a certain number of degrees about a certain point. Rotation problems are not included on the test.

Practice Problems

1. Twice the supplement of the angle equals 5 times the complement of the angle. What does the angle measure?
   (a) ninety degrees
   (b) sixty degrees
   (c) thirty degrees
   (d) fifteen degrees
2. What is the midpoint of the line segment connecting(−8, −4) and (2,6)?
   (a) (−3,−1)
   (b) (3,−1)
   (c) (3,1)
   (d) (−3,1)
3. What are the coordinates of the point (−5,7) after it is reflected over the x-axis and then translated two units to the right and four units down?
   (a) (7, 3)
   (b) (−3,−11)
   (c) (−3,3)
   (d) (3,11)
4. What are the coordinates of the point (4,3) after it is reflected over the line $y = x$ and then shifted three units to the left and four units up?
   (a) (7,7)
   (b) (6,0)
   (c) (5,1)
   (d) (0,8)

Detailed answers for above problems:

1. Answer: (c) thirty degrees
   Always begin a word problem by defining the variable. In this case, the problem is asking for the measure of the angle, so let x = the measure of the angle. The supplement of an angle is 180 minus the angle, and the complement is 90 minus the angle. Therefore, twice the supplement can be represented as 2(180−x), while five times the complement equals 5(90−x). Thus, 2(180−x)=5(90−x). Using the distributive property, 360−2x=450−5x. Add 5x to both sides, so that 360+3x=450. Now, subtract 360 from both sides, giving 3x=90. Dividing both sides by 3 results in the final answer of thirty degrees.
2. Answer: (d) (−3,1)
   To find the midpoint of the line segment connecting two points, use the midpoint formula. Let the point (−8, −4) equal (x_1, y_1) and the point (2, 6) equal (x_2, y_2). Now, plug into the formula $(\frac{x_1+x_2}{2}, \frac{y_1+y_2}{2})$. Therefore, the midpoint equals $\frac{-8+2}{2}, \frac{-4+6}{2}$ or $(\frac{-6}{2}, \frac{2}{2})$ = (−3, 1).
3. Answer: (b) (−3,−11)

Measurement and Geometry

When a point is reflected over the x-axis, the x value in the ordered pair remains the same, but the value for y changes its sign. The reflected point would be $(-5,-7)$. Now shifting the point two units to the right and four units down can be found by adding 2 to the x coordinate and subtracting 4 from the y coordinate: $-5+2, -7-4 = (-3, -11)$.

4. Answer: (d) (0,8)

   When reflecting over the line $y = x$, simply interchange the values of x and y. In other words, the new point is (y,x). Thus, the new point is $(3,4)$. Now shifting three units to the left means subtracting 3 from the x-coordinate, while moving up four units means adding 4 to the y-coordinate. $(3-3, 4+4) = (0, 8)$ is the final result.

Chapter Six

Statistics and Probability

MEASURES OF CENTRAL TENDENCY AND SPREAD

When looking at large groups of numbers it is important to organize the numbers so that it is possible to draw conclusions or make predictions based on the data.

Measures of Central Tendency are more commonly known as mean, median, and mode. These are three measures that give information about groups of numbers.

The *mean*, or average, is the number found by adding up all the numbers in the given set of numbers and then dividing that sum by the total number of numbers. The calculated mean may not be a number that was in the original set of numbers but it will be a number that is between the smallest and the largest number in the set of numbers.

Example 1

Find the mean of 30, 40, 15, 44, and 62
    Answer: 38.2

$$\frac{30 + 40 + 15 + 44 + 62}{5} = \frac{191}{5} = 38.2$$

The *median* is the number in the middle of a group of numbers arranged from smallest to largest. The median can be found by writing all the numbers from smallest to largest and then crossing off a number at either end of the list. Keep crossing off numbers on each end and work toward the middle. The number that is in the middle is the median. If there is an even number of numbers in the set, then there will be two numbers in the middle. Add these two numbers together and divide that sum by 2 to find the median.

Example 1

Find the median of 30, 62, 40, 53, and 44

$$\cancel{30}, \cancel{40}, 44, \cancel{53}, \cancel{62}$$

   Answer: 44 is the median

Example 2

Find the median of 121, 164, 155, 54, 543, and 432

$$\cancel{54}, \cancel{121}, 155, 164, \cancel{432}, \cancel{543}$$

Answer: 159.5

$$\frac{155+164}{2} = \frac{319}{2} = 159.5$$

Another way to find the median is to add 1 to the number of numbers and then divide that sum by 2. That number will be the middle number. Count over that number of numbers in the ordered list and that is the median. If the calculated number is not a whole number, then use the whole numbers on either side of that number, add them together, and divide that sum by 2.

Example 3

Find the median of 30, 62, 40, 53, and 44
Answer: 44
There are five numbers given. $\frac{5+1}{2} = \frac{6}{2} = 3$. Count over to the 3rd number and that is the median.

$$30, 40, \boxed{44}, 53, 62$$

Example 4

Find the median of 121, 164, 155, 54, 543, and 432
Answer: 159.5
There are six numbers given. $\frac{6+1}{2} = \frac{7}{2} = 3.5$. Find the 3rd number (155) and the 4th number (164).

$$54, 121, \boxed{155}, \boxed{164}, 432, 543$$

$$\frac{155+164}{2} = \frac{319}{2} = 159.5$$

The *mode* is the number that appears most often in a set of numbers. A way to remember this term is by using the acronym **M**ost **O**ften **D**oes **E**xist.

Example 1

Find the mode of 23, 65, 43, 23, and 57
Answer: 23
There are two number 23's, so 23 is the mode.

Example 2

Find the mode of 3, 5, 6, 3, 6, 10, and 21
   Answer: 3, 6
   There are two number 3's and two number 6's, so there are two modes: 3 and 6.

Example 3

Find the mode of 10, 20, 30, 40, and 50
   Answer: no mode
   There is not more of one number than any other number, so this set of numbers has no mode.

Spread, or *range*, is a way of looking at how much variation there is in a group of numbers. The range of a group of numbers is found by subtracting the smallest number from the largest number.

Example 1

Find the range of 34, 76, 27, 98, and 22
   Answer: 76

$$98 - 22 = 76$$

Practice Test Questions on Measures of Central Tendency and Spread

1. A student needs to earn an average of 80 to get a B in English class. So far, she has earned scores of 76, 82, 70, and 88. What is the minimum score that she needs to get on her last test to earn a B in the class?
   (a) 79
   (b) 84
   (c) 80
   (d) 94
2. Several houses in Harleysville have been sold over the past six months for the following prices: $312,000, $296,000, $320,000, $259,600, $300,700, and $324,500. What was the median selling price?
   (a) $289,900
   (b) $312,000
   (c) $102,117
   (d) $306,350
3. What is the mode or modes of the following list of numbers?
   0, 1, 6, 5, 1, 2, 1, 2, 8, 2, 0
   (a) 1 and 2
   (b) 1
   (c) 2
   (d) 0, 1, and 2

4. In a small village in 1850, there were a 100 families with a yearly income of $300. One lucky and hardworking family had a yearly income of $2,000. Which measure of central tendency most accurately represents the yearly income of a family in this village?
   (a) mean
   (b) median
   (c) mode
   (d) median or mode

Detailed answer key for the above problems:

1. Answer: (b) 84
   So far, she has earned a total of 76 + 82 +70 + 88 = 316 points. Using the formula to find the mean, all the test scores are added together (316 + the last test) and then divided by the number of tests she has taken (5) to get a mean score of 80.
   $\frac{316+x}{5}=80$.
   Multiply both sides of the equation by 5. This results in 316 + x = 400. Subtract 316 from both sides of the equation. This gives x = 84.
2. Answer: (d) $306,350
   First, the selling prices of the homes need to be listed from least to greatest:
   259,100, 296,000, 300,700, 312,000, 320,200, 324,500
   There are two prices in the middle: 300,700 and 312,000. These two prices in the middle are added together to get the sum of 612,700, and then this total is divided by 2 to get the answer: $306,350.
3. Answer: (a) 1 and 2
   The number 1 occurs three times and the number 2 also occurs three times. There are only two 0's. So 1 and 2 are the mode.
4. Answer: (d) median or mode
   The mean would be $\frac{100 \cdot 300+2000}{101}$, which is about $317 dollars. Although this is the mean of the yearly incomes, it does not accurately reflect the fact that all but one family has an income of $300. The median and the mode are both $300, which is a more accurate representation of the yearly income in the village.

DISPLAYING DATA AND STATISTICAL INFORMATION

When presenting data, a list of numbers, it needs to be organized so that the data can be used to draw conclusions.

A *table* can be used to present a list of data. The data is presented in columns with headings.

This table can now be used to answer questions about the data.

Rainfall in Cities in the Months of April and June

City	April	June
New York, NY	4 inches	4 inches
Pittsburgh, PA	3.5 inches	4 inches
Miami, FL	3 inches	1.25 inches
Phoenix, AZ	0.8 inches	0.25 inches

Example 1

What city has the most rain in the month of April?
  Answer: New York, NY
  Look at the April column, find the biggest number (they are all measurements in inches), which is 4 inches, and find the city that goes with that number.

A *circle graph* or *pie chart* can be used to show the relationship of a category of data to the total of all categories. Each section of the circle is a percentage of the whole.

Example 1

Create a pie chart representing favorite ice cream flavors using these data.
  Answer:

Favorite Ice Cream Flavor

Vanilla	6 people
Chocolate	4 people
Rocky Road	10 people
Total	20 people

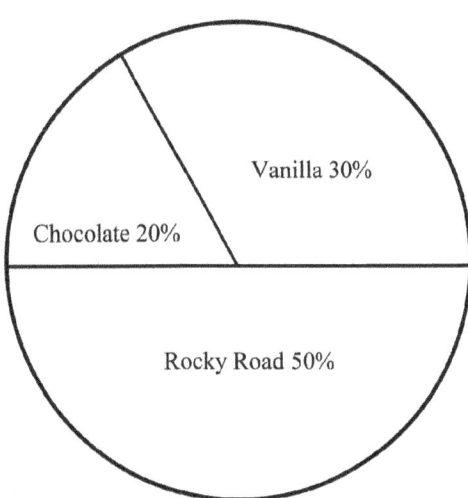

Rocky Road is the favorite of half of the people.
To create the pie chart:

1. Change the number of people into a percentage.
   (a) Vanilla $\frac{6}{20}=30\%$
   (b) Chocolate $\frac{4}{20}=20\%$
   (c) Rocky Road $\frac{10}{20}=50\%$

2. Draw the pie chart and label it with the ice cream flavor and its percentage. Half of the circle is Rocky Road, so draw a line cutting the circle in half. Label the bottom of the circle "Rocky Road." The top half of the circle is 50% of the circle. Use tick marks around the top half to cut it into five equal pieces—each piece is 10%. Now draw a line at 20%. That piece is Chocolate and the remaining 30% is Vanilla.

A *histogram* is used to show data that can be grouped into intervals. The frequency of each interval is then represented in the graph with bars that are the same width. These bars are touching because the intervals are continuous, which means there are no breaks between the intervals.

Example of a Histogram

Test scores on an algebra test:
   Data: 52, 72, 73, 73, 78, 79, 79, 80, 81, 82, 83, 84, 85, 86, 88, 90, 95, 99

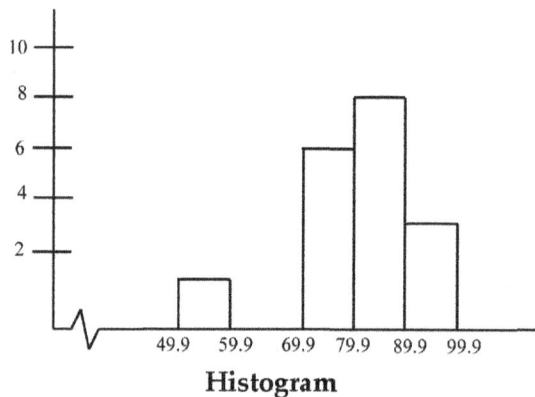

Histogram

Interval	Frequency
50 – 59.9	|
60 – 69.9	
70 – 79.9	𝃪 |
80 – 89.9	𝃪 |||
90 – 99.9	|||

Frequency Chart

A *stem-and-leaf plot* is a way to show how often data values appear in an interval. The data is summarized, but the individual values are not lost. To create this plot, the last digit of a number is the leaf and the rest of the number is the stem. A key is always part of a stem-and-leaf plot.

Example of a Stem-and-Leaf Plot

Data: 111, 131, 120, 124, 124, 152, 133, 137, 138, 126, 140, 140, 140, 116, 127

Stem	Leaf
11	1 6
12	0 4 4 6 7
13	1 3 7 8
14	0 0 0
15	2

Key 14 | 0 = 140

Stem-and-Leaf Plot

Example of Histogram for the Same Data

Interval	Frequency
100 – 119.9	||
120 – 129.9	ﾊﾄ (5)
130 – 139.9	||||
140 – 149.9	|||
150 – 159.9	|

116   *Chapter Six*

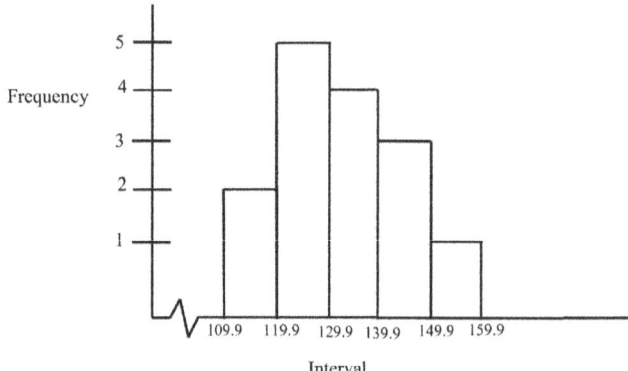

In both graphics it is easy to see there are more numbers in the 120–129.9 interval. The stem-and-leaf plot shows the longest list of numbers and the histogram shows the highest bar in that interval. However, the stem-and-leaf plot is the one where the individual numbers are maintained; 137 and 124 are still visible. The histogram loses the exact numbers and just shows the interval of data.

A *bar graph* is another way to show the frequency of data with a picture. Bar graphs are best when there are categories of data rather than numerical data.

Example of a Bar Graph

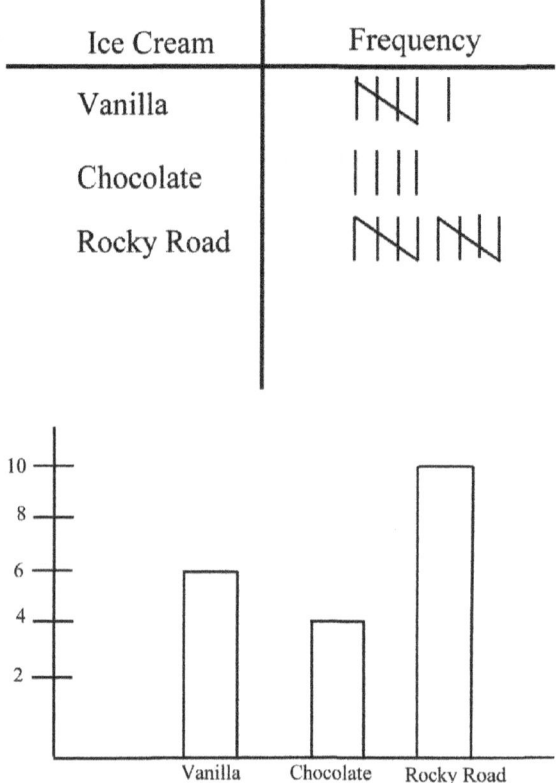

A *box-and-whiskers plot* is a graph that displays statistical information that shows how data is spread between its upper and lower values. The data will be summarized to five numbers and these values will be graphed. To create this plot, the median of the data is calculated. This is one of the five points that will be graphed. Then, the median of the lower half of the data is calculated (first quartile) and the median of the upper half of the data is calculated (third quartile). These are the second and third values that are graphed and are the left and right sides of the rectangle or box. The lowest data value (minimum) and the highest data value (maximum) are then plotted as dots with a line from the dot to the rectangle.

Example of a Box-and-Whiskers Plot

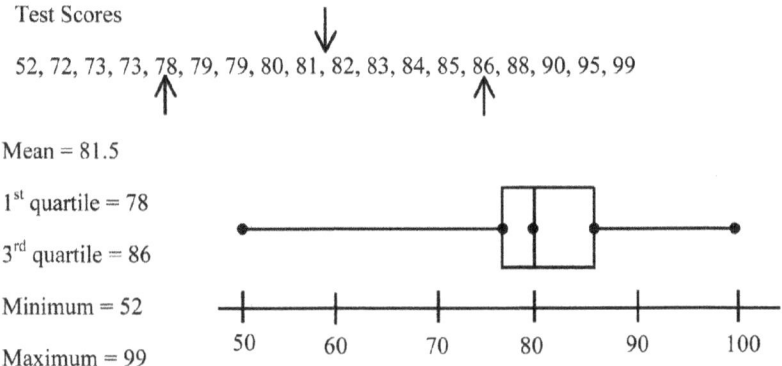

The length of the rectangle shows how closely the center half of the data is grouped around the median or middle values.

Practice Test Questions on Displaying Data and Statistical Information

1. This is an example of which type of graph?

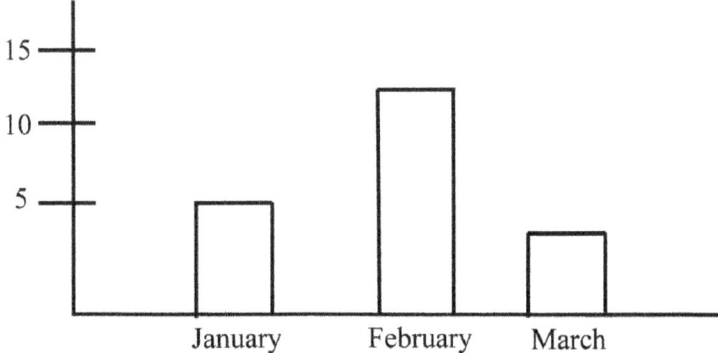

   (a) Histogram
   (b) Bar graph
   (c) Box-and-whiskers plot
   (d) Stem-and-leaf plot

2. What is the median weight of the wrestling team?

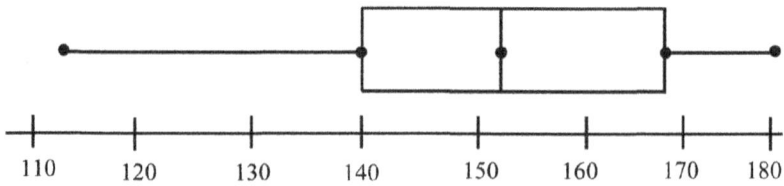

   (a) 135
   (b) 150
   (c) 152
   (d) 160
3. The type of graphs that most clearly displays the lowest and highest data value is
   (a) Histogram and bar graph
   (b) Box-and-whisker plot and bar graph
   (c) Stem-and-leaf plot and histogram
   (d) Box-and-whisker plot and stem-and-leaf plot
4. Which type of graph shows the frequency of data?
   i. Histogram
   ii. Bar graph
   iii. Box-and-whisker plot
   iv. Stem-and-leaf plot
      (a) i, ii, and iv
      (b) iii
      (c) i and ii
      (d) i, ii, iii, and iv

Detailed answer key for the above problems:

1. Answer: (b) bar graph
   The bars are not continuous—they don't touch each other—so it is not a histogram. Individual data values are not displayed so it is not a stem-and-leaf plot. The graph does not have a box or individual points displayed so it is not a box-and-whisker plot.
2. Answer: (c) 152
   The line in the middle of the box is the center of the data. The box goes between 140 and 170 so that eliminates the answer of 135. The center of the box is between 150 and 160, which then points to the value of 152.
3. Answer: (d) box-and-whisker plot and stem-and-leaf plot
   Both of these display individual data values.
4. Answer: (a) i, ii, and iv
   The box-and-whisker plot is the only type of graph that does not display individual data values.

ANALYZING AND DRAWING INFERENCES FROM DATA PRESENTED IN DIFFERENT FORMATS

A *frequency distribution* is a table that organizes numerical data into intervals and shows how often (the frequency) that the data falls into each interval.

Example of a Frequency Distribution

Looking at the following table, these conclusions can be drawn:

Height (in inches)	Tally	Frequency
45 – 49.9	|	1
50 – 54.9	|||	3
55 – 59.9	︴	5
60 – 64.9	||	2
65 – 69.9	|	1

1. There is one student that is 45–49.9 inches tall.
2. The interval with the greatest number of students is 55–59.9.

A *percentile* is a statistical measure related to percentages. It is often used with standardized tests and is a way of determining how one score relates to the scores of everyone else that took the test. A percentile is a single number. A score in the 72nd percentile means that approximately 72% of the data lie at or below the given data.

A *graph* is a picture view of data. Graphs can be in the shape of a circle, a coordinate plane, or a number line. Pie charts, bar graphs, histograms, stem-and-leaf plots, and box-and-whisker plots are all examples of graphs.

Practice Test Questions on Analyzing and Drawing Inferences from Data

1. How many people responded to the questionnaire about their muffin preference?
   (a) 15
   (b) 1600
   (c) 1400
   (d) 14

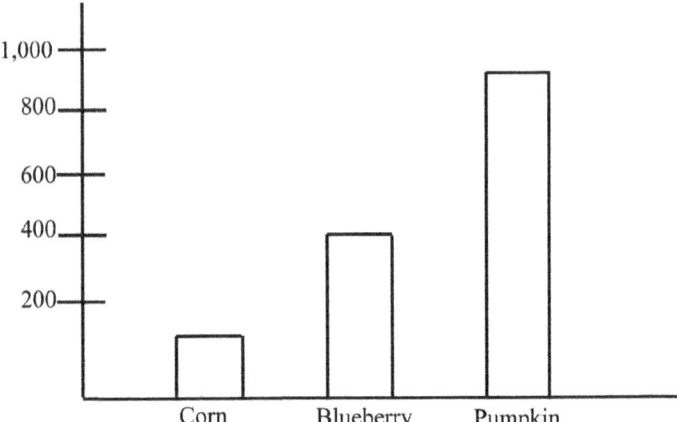

2. Bridget scored in the 83rd percentile on a test. She scored below what percentage of the test-takers?
   (a) 17%
   (b) 82%
   (c) 75%
   (d) 20%
3. The principal of Cabrini Elementary School prepares a report for the school board showing the top third-grade scores on the state reading test. Two of the three figures in the report reflect the same data. Which two are they?
   (a) i and ii
   (b) i and iii
   (c) ii and iii
   (d) all three show the same thing

Interval	Frequency
70 – 79.9	60
80 – 89.9	40
90 – 99.9	15

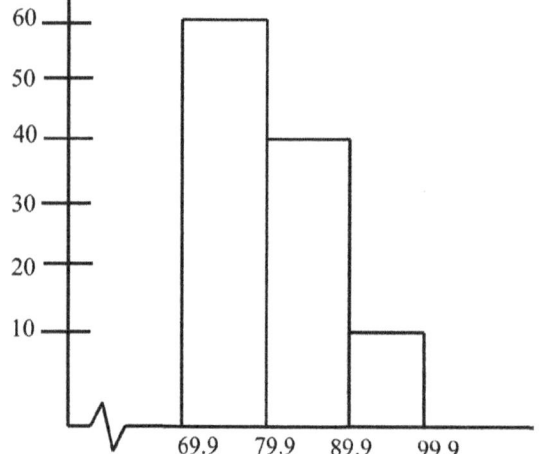

Percentile	Number of Students
70 – 79.9	60
80 – 89.9	40
90 – 99.9	15

4. What activity or activities take up 6 hours of the student's day?
   (a) Watch TV
   (b) Homework and job
   (c) Eating
   (d) Job

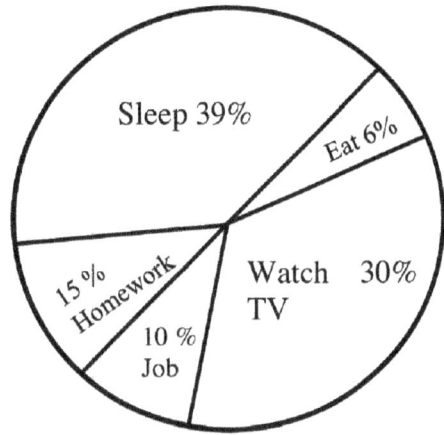

Detailed answer key for the above problems:

1. Answer: (c) 1400
   One hundred people like corn muffins, 400 hundred people like blueberry muffins, and 900 hundred people like pumpkin muffins. That is a total of 1,400 people.
2. Answer: (a) 17%
   Bridget scored above 83% of the test-takers. 100% − 83% = 17%. 17% of the scores were above the student.
3. Answer: (a) i and ii
   The frequency chart and the histogram are reporting actual scores. A percentile is showing a comparison of one student's score to the rest of the students that took the test.
4. Answer: (b) homework and job. Homework is 15% and job is 10%
   This is a total of 25% or $\frac{1}{4}$ of the day. $\frac{1}{4} \times 24$ hours = 6 hours.

PROBABILITY: SIMPLE, COMPOUND, INDEPENDENT, DEPENDENT, CONDITIONAL

A *set* is a group of related elements. Sets are used in calculating probabilities. *Probability* is used to decide how likely it is that something will or will not happen. The outcomes must all be equally likely to happen. The denominator of a probability is the set of all the equally likely possible outcomes. The numerator is a subset—the part of the set in the denominator that answers what is being asked in the question.

Equally Likely	Not Equally Likely
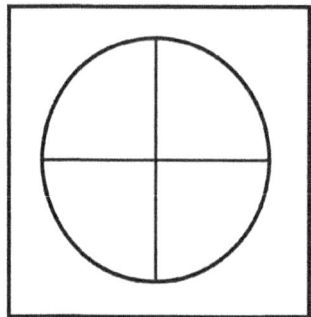	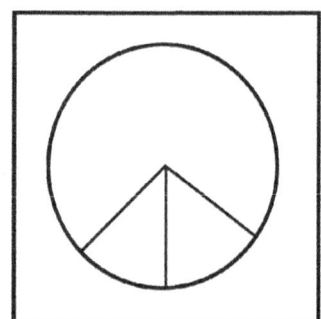

Probability is found using this formula:

$$\frac{\text{Number of possible successes}}{\text{Total number of outcomes}} \text{ or } \frac{\text{What does the problem ask}}{\text{Everything that could happen}}$$

Probabilities are fractions or decimals that equal a number between 0 and 1, inclusive. If there is no chance of an event happening, the probability is zero. For example, the probability of rolling a 7 with one die is 0. If the event is certain to happen, its probability is 1. The probability of rolling a 1, 2, 3, 4, 5, or 6 with one die is 1. All other probabilities will fall between 0 and 1.

Simple probability is the probability of one event happening.

Example 1

What is the probability of rolling a 2 with one die?
    Answer: $\frac{1}{6}$

$$\frac{1 \text{ way to roll a 2}}{6 \text{ possible numbers on a die}} = \frac{1}{6}$$

Example 2

There are 2 yellow, 5 green, and 6 blue balloons on the floor. What is the probability that a green balloon will pop?
    Answer: $\frac{5}{13}$

$$\frac{\text{There are 5 green balloon}}{\text{There are a total of 13 balloons}}$$

So the probability is $\frac{5}{13}$ that a green balloon will pop.

Probability can be found using one or more related sets. Sometimes the same element can be in 2 or more sets. This means the sets overlap or intersect.

Example of an Intersection of Two Sets

Set A = {girls in English class}
Set B = {education majors in English class}

The intersection of these two sets would be the female education majors in English class. This problem can be written as A∩B. *Intersection* means "and." The elements have to be in set A and in set B.

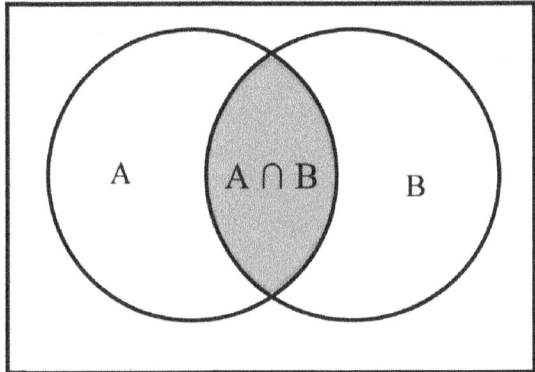

A∩B has been shaded

Example of a Union of Two Sets

A *union* of two or more sets means that the element can be in one set, or in the other set, or in both sets. This is written as A∪B.

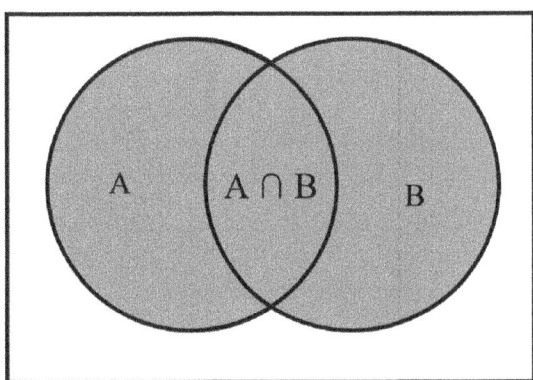

A∪B has been shaded.

Example 3

There are 84 students headed to Florida for Spring Break; 51 are planning to rent beach umbrellas, 30 students are bringing coolers, and 17 will have a cooler and rent a beach umbrella. Fill in a Venn Diagram showing this information.

Answer:

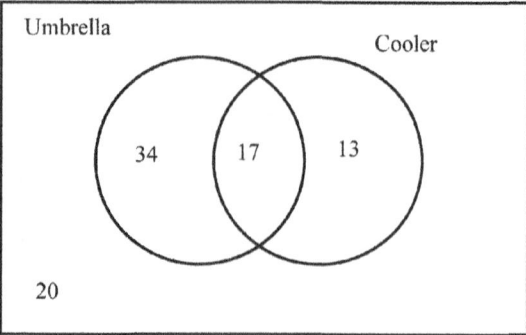

1. Decide how many circles should be in the Venn Diagram. A circle is needed for renting beach umbrellas and a circle is needed for coolers so two circles are drawn.

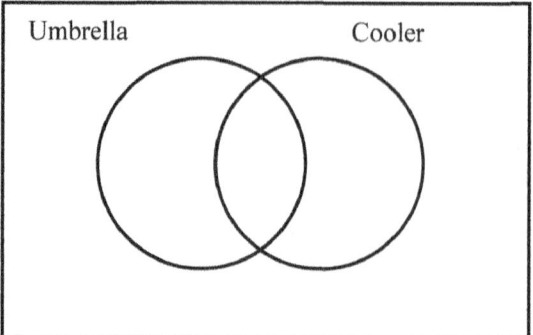

2. Fill in the intersection of the two circles first—the region where the circles overlap, region A∩B. This information is given in the statement "17 will have a cooler and rent a beach umbrella." Put a 17 in the overlapping region.

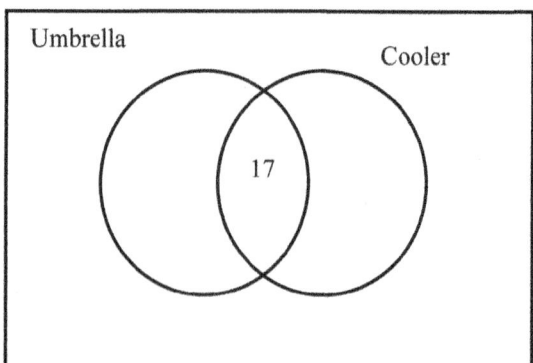

3. The problem states that 51 students will rent beach umbrellas so the entire beach umbrella circle should contain fifty-one students. Right now the circle for beach umbrellas already has 17 students in it so these students need to be removed from the 51 total students. 51−17=34. Put 34 more students in the circle by putting 34 in region A.

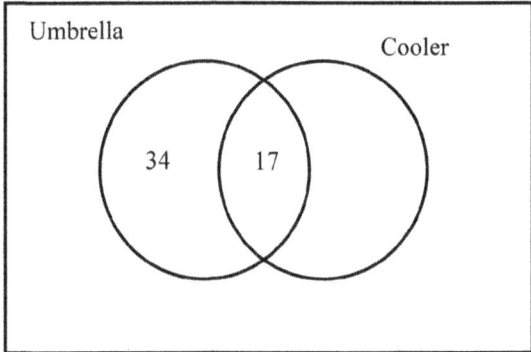

4. The problem states that 30 students will bring coolers. Following the same logic as step 2, the entire cooler circle should contain 30 students. Right now the circle for coolers already has 17 students in it so these students need to be removed from the 30 total students. 30−17=13. Put 13 more students in the circle by putting 13 in region B.

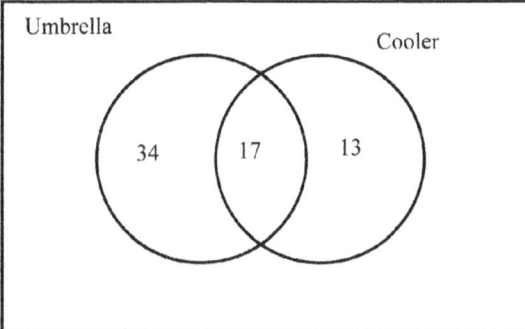

5. The problem states that 84 students were heading to Florida. So, all 84 students must be represented in the Venn Diagram. So far there are 34+17+13=64 students in the diagram. The other twenty students, 84−64=20, go outside of both circles.

Example 4

Using the Spring Break Venn Diagram from Example 2, answer this question: What is the probability a student will rent an umbrella or have a cooler?

Answer: 0.76

Using the numbers from the Venn Diagram, add up the numbers inside the circles and divide that total by all of the students:

$$\frac{34 + 17 + 13}{84} = \frac{64}{84} = \frac{16}{21} \approx 0.76$$

Be sure not to count the students renting an umbrella and using a cooler twice.

51 renters + 30 with coolers − 17 doing both = 64 students
n(renters) + *n*(coolers) − *n*(renters ∩ coolers) = *n*(renters ∪ coolers)

Example 5

Using the Spring Break Venn Diagram from Example 3, answer this question: What is the probability a student will rent an umbrella and have a cooler?
Answer: 0.2
Using the numbers from the Venn Diagram, use the number in the intersection of renting an umbrella and using a cooler:

$$\frac{17}{84} \approx 0.2$$

Compound probability is the probability of more than one event happening. To find this probability, the first step is to list all the possible outcomes. This is often done by making a tree diagram. This set of outcomes is called the sample space and is the denominator of the fraction. After all the outcomes are listed, the next step is finding the outcomes that meet the criteria asked for in the question.

Example 1

A family plans on having three children. What is the probability that the children in the family picture will be two girls and one boy?
Answer: $\frac{3}{8}$

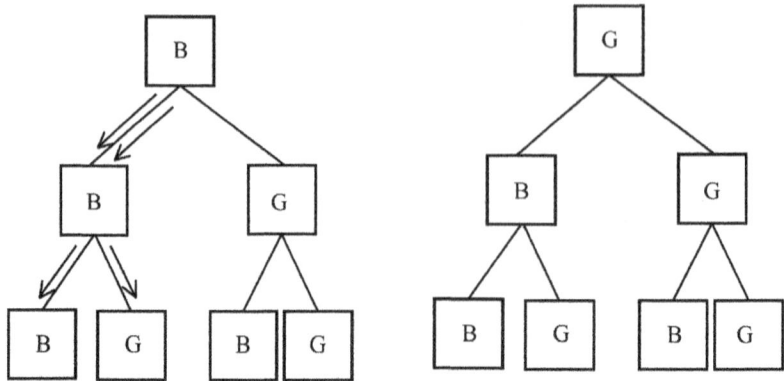

Follow each branch of the tree and list the outcomes.

{BBB, BBG, BGB, BGG, GBB, GBG, GGB, GGG}

There are 8 different branches in the tree diagram and there are 8 outcomes.
Now, count the outcomes that include 2 girls. These are {BGG, GBG, GGB} There are 3 successful branches. So, the probability of 2 girls and 1 boy in the family picture is $\frac{3}{8}$.

When finding the probability of more than one event, there are two rules to keep in mind. To find the probability of one event and another event happening, multiply the probability of each event together.

When finding the probability of one event or another event, add the probability of each event together, being careful not to double count any of the outcomes.

In Example 1, finding the probability of having two girls and one boy could be calculated this way: The probability of having a girl is $\frac{1}{2}$. The probability of having a boy is $\frac{1}{2}$. So, the probability of having 2 girls and a boy is $\frac{1}{2} \times \frac{1}{2} \times \frac{1}{2} = \frac{1}{8}$ and there are three different ways this can happen—BGG or GBG or GGB—so $\frac{1}{8} + \frac{1}{8} + \frac{1}{8} = \frac{3}{8}$.

Independent probability means that one event has no effect on a second event. One example of this is flipping a coin more than once. The outcome of getting a head the first time does not change the outcome of getting a head the second time. When more than one independent event happens, the probabilities of the individual events are multiplied together.

Example 1

Find the probability of drawing a heart, putting the card back in the deck and drawing another heart.
   Answer: $\frac{1}{16}$

There are 13 hearts in a deck and 52 cards in a deck of cards. So, the probability of drawing the first heart is $\frac{13}{52}$ or $\frac{1}{4}$. Then, the card is put back and another card is drawn. There are still 13 hearts and 52 cards so the probability of drawing the second heart is $\frac{13}{52}$ or $\frac{1}{4}$. Now these 2 events are put together using the word *and*. A heart and a heart are necessary to meet the requirement of the question. The word *and* means the probabilities of the events are multiplied together.

$$\frac{1}{4} \times \frac{1}{4} = \frac{1}{16}$$

The probability of drawing 2 hearts, with replacement, is $\frac{1}{16}$.

Dependent probability means that the first event changes the probability of a subsequent event. The previous example can be changed slightly to illustrate this. Instead of drawing a card, replacing it, and drawing another card, the first card will not be replaced.

Example 1

What is the probability of drawing two hearts without replacement?
   Answer: $\frac{1}{17}$

Now, the probability of drawing the second heart will change. One heart has been removed from the deck. So the probability of drawing the second card will be calculated like this: There are now only 12 hearts in the deck of cards and there are only 51 cards. So the probability of drawing the second heart is $\frac{12}{51}$. To find the probability of drawing a heart, keeping it, and drawing another card is $\frac{1}{4} \times \frac{12}{51} = \frac{12}{204} = \frac{1}{17}$

Conditional probability means that only part of a set of outcomes will be considered in finding the probability of an event. This type of question often refers to charts or Venn Diagrams. The number in the denominator is the number of outcomes of the given condition. The probability of the second event depends on the first event having already happened.

Example 1

What is the probability that a person is a runner, given that she is a woman?
Answer: $\frac{13}{78} = \frac{1}{6}$

	Men	Women	Totals
Runners	7	13	20
Nonrunners	35	65	100
Totals	42	78	120

This question is only concerned with women so the first and third columns are irrelevant; cross them out. Now, looking only at the second column, there are 13 runners out of 78 total women. The probability that a person is a runner, given that she is a woman, is $\frac{13}{78} = \frac{1}{6}$

Example 2

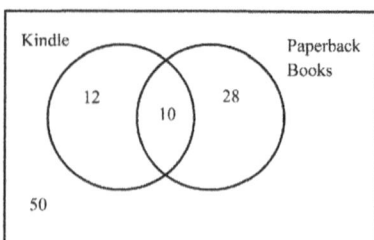

What is the probability that Diane reads a book on her Kindle, given that she reads paperbacks?
Answer: $\frac{5}{19}$

Only look at the paperback circle because that is the given event. There are 38 books in that circle, and this is the number in the denominator. Now, of those 38 books, ten of them are reading using a Kindle. That is the numerator. So the probability that Diane will read a book on her Kindle, given that she reads paperbacks, is $\frac{10}{38} = \frac{5}{19}$

Practice Test Questions on Probability: Simple, Compound, Independent, Dependent, Conditional

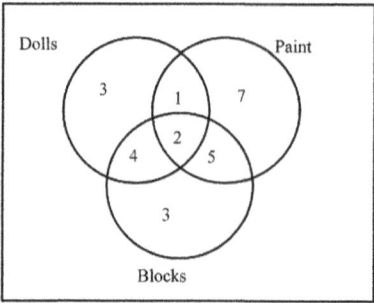

1. Students in a pre-school class were given time to visit three centers in their classroom. Some played at one center while others played at more than one center. What is the probability that a student visited more than one center?
   (a) $\frac{10}{25}$
   (b) $\frac{11}{25}$
   (c) $\frac{12}{25}$
   (d) $\frac{13}{25}$

2. The Harleysville Motorcycle Club has 340 members; 272 of the members are men. One hundred of the members are overweight. Thirty of the women are overweight. What is the probability that a motorcycle club member will be a man, given that the member is not overweight?
   (a) $\frac{202}{240}$
   (b) $\frac{272}{340}$
   (c) $\frac{240}{340}$
   (d) $\frac{202}{272}$

3. There are 3 blue marbles, 5 red marbles, and 8 yellow marbles in a bag. One marble is selected and put back in the bag, and then a second marble is selected. What is the probability that two blue marbles are drawn?
   (a) $\frac{3}{16} + \frac{3}{16}$
   (b) $\frac{3}{16}$
   (c) $\frac{3}{16} \times \frac{2}{15}$
   (d) $\frac{3}{16} \times \frac{3}{16}$

4. A coin is tossed and a fair die is rolled. What is the probability of getting a tail and an even number?
   (a) $\frac{1}{2}$
   (b) $\frac{1}{4}$
   (c) 0
   (d) $\frac{1}{8}$

Detailed answer key for the above problems:

1. Answer: (c) $\frac{12}{25}$
   The denominator is the twenty-five students represented in the Venn Diagram on page 128. The numerator is the students that used more than one center. These students are in the areas where the circles overlap: 1+2+4+5=12.

2. Answer: (a) $\frac{202}{240}$
   Make a chart. Start with the total of 340 members in the bottom right corner. Then, fill in 272 men in the bottom row. The total number of women is calculated by subtracting 272 from 340. 340−272=68 women. Next, fill in 100 overweight people in the last

column. Find the not-overweight people by subtracting 100 from 340, giving a total of 240 not-overweight people. The right column and the bottom row are now complete. The problem stated that there are 30 overweight women. This number goes in the second box across the top. The first box across the top, overweight men, is calculated by subtracting 30 from 100, which is 70. Finish the first column by 272−70=202. The middle column is completed by 68−30=38.

	Men	Women	Total
Overweight	70	30	100
Not Overweight	202	38	240
Total	272	68	340

The given is that the person is not overweight, so only look at the second row. The males in that row number 202, which is the numerator. The denominator is the total number of not-overweight people and is found in the third column, which is 240.

3. Answer: (b) $\frac{3}{16} \times \frac{3}{16}$

   The probability of selecting a blue marble is $\frac{3}{16}$. These two events are independent so the probability of the second event is also $\frac{3}{16}$. Then, these probabilities are multiplied together.

4. Answer: (b) $\frac{1}{4}$

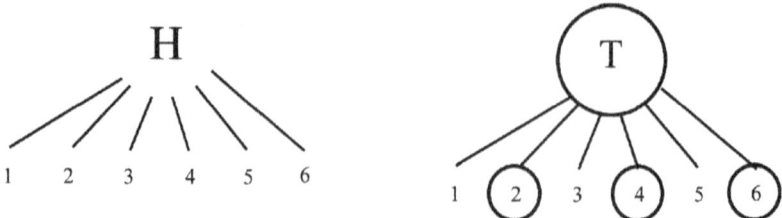

There are twelve possible outcomes. Three of the outcomes are the event of getting a tail and an even number. So $\frac{3}{12} = \frac{1}{4}$. Another way of calculating the answer is to find the probability of each independent event and multiply them together. The probability of getting a tail is $\frac{1}{2}$. The probability of rolling an even number is $\frac{3}{6} = \frac{1}{2}$, so $\frac{1}{2} \times \frac{1}{2} = \frac{1}{4}$.

COUNTING PRINCIPLES, PERMUTATIONS AND COMBINATIONS

The Fundamental Counting Principle, also called the Multiplication Principle, states that if there are *m* ways for one event to occur, and *n* ways for a second event to occur, then there are $m \times n$ ways for both to occur. So, this is used to find out how many ways different choices can be put together.

Example 1

How many different outfits of a sweater, pants, and shoes can be put together if there are 3 different sweaters, 4 different pairs of pants, and 2 different pairs of shoes from which to choose?

Answer: 24 outfits

A tree diagram could be used to find how many different outfits of a sweater, pants, and shoes can be put together if there are 3 different sweaters, 4 different pairs of pants, and 2 different pairs of shoes from which to choose.

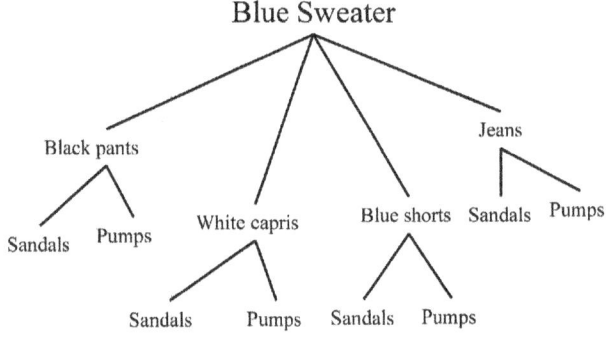

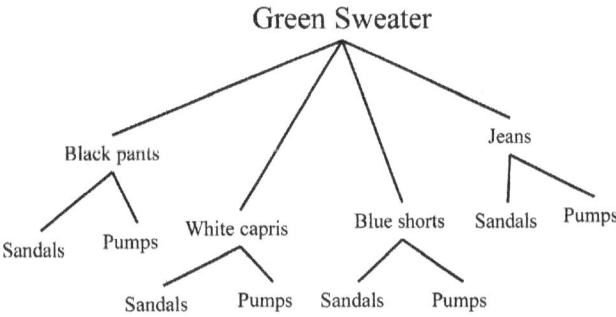

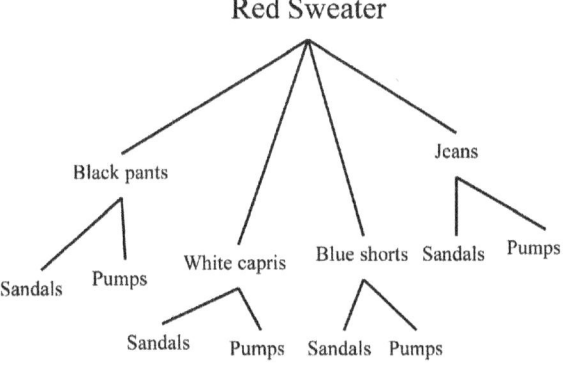

One outfit might be a blue sweater, jeans, and sandals. By counting along the bottom of each branch, there are twenty-four different possible outfits that can be put together.

To avoid having to draw tree diagrams, the Fundamental Counting Principle states that the numbers of each choice are multiplied together to find the total number of possibilities.

$$\underset{\text{sweaters}}{3} \times \underset{\text{pants}}{4} \times \underset{\text{shoes}}{2} = 24$$

A *permutation* is an arrangement of objects. The order of the objects is important. Reading a math book, then a history book, and then a novel is different from reading a novel, then a math book, and then a history book. Permutations can be written as $P(n,r)$ or $_nP_r$. The formula for calculating a permutation is $\frac{n!}{(n-r)!}$ where n is the number of objects to choose from and r is the number of objects that are being chosen. ! is the factorial symbol. It is an abbreviation for a special type of multiplication. Start with the number before the factorial symbol and multiply it by every number found when counting down to 1. For example, $4! = 4 \times 3 \times 2 \times 1 = 24$.

To solve a permutation of 3 objects taken 2 at a time using the permutation formula, it looks like this: $P(3,2) = \frac{3!}{(3-2)!} = \frac{3!}{1!} = \frac{3 \times 2 \times 1}{1} = 6$

The calculator key $_nP_r$ can also be used.

Example 1

Margie is going on a trip and is taking 6 CD's along. How many different ways can she listen to 4 of those CD's?

Answer: 360 different ways

She has 6 CD's and is taking 4 at a time. The order she listens to them is important. Listening to the Beatles before the Rolling Stones is different from listening to the Rolling Stones and then the Beatles. So, $P(6,4) = \frac{6!}{(6-4)!} = \frac{6!}{2!} = \frac{6 \times 5 \times 4 \times 3 \times 2 \times 1}{2 \times 1} = 6 \times 5 \times 4 \times 3 = 360$.

Margie can listen to her 4 CD's in any of 360 different orders.

A *combination* is a way of grouping objects. The order the objects are selected is not important. If books are selected for a library used-book sale, it doesn't matter what order they are put in a pile because the whole pile is going to the library. Combinations are written as $C(n,r)$ or $\binom{n}{r}$ or $_nC_r$. The formula for calculating a combination is $\frac{n!}{(n-r)!r!}$ where n is the number of objects to choose from and r is the number of objects that are being chosen. To solve a combination of 3 objects taken 2 at a time using the combination formula, it looks like this: $C(3,2) = \frac{3!}{(3-2)!2!} = \frac{3!}{1!2!} = \frac{3 \times 2 \times 1}{1 \times 2 \times 1} = \frac{6}{2} = 3$.

The calculator key $_nC_r$ can also be used.

Example 1

A group of 20 students needs to select a committee of 3 to plan their Spring Break Habitat for Humanity trip. How many different committees are possible?

Answer: 1,140 committees

$$C(20,3) = \frac{20!}{(20-3)!\,3!} = \frac{20!}{17!\,3!} = \frac{20 \times 19 \times 18 \times 17!}{3 \times 2 \times 1 \times 17!} = \frac{20 \times 19 \times 18}{3 \times 2 \times 1} = \frac{6840}{6} = 1140$$

There are 1,140 different committees that could be formed.

Combinations can get more complicated when specific items are being chosen from a group of objects. Then, a method of organization is needed. A have/want chart meets this need.

Example 2

There are 8 seniors, 4 juniors, 3 sophomores, and 5 freshmen in a writing class. Two students from each class year are picked to go to a writing conference. How many different groups of eight students could be selected?

Answer: 5,040 different groups of students

First, record the specific column headings for the categories being chosen. Then, write down what there is to select from—the "have." Then, underneath that, write down what is needed—the "want."

	Seniors	Juniors	Sophomores	Freshmen
Have	8	4	3	5
Want	2	2	2	2

Each column is a combination. Then, because seniors *and* juniors *and* sophomores *and* freshmen are needed, the Fundamental Counting Principle is used and the combinations are multiplied together.

$$C(8,2) \times C(4,2) \times C(3,2) \times C(5,2)$$

$$\frac{8!}{(8-2)!\,2!} \times \frac{4!}{(4-2)!\,2!} \times \frac{3!}{(3-2)!\,2!} \times \frac{5!}{(5-2)!\,2!} = \frac{8!}{6!\,2!} \times \frac{4!}{2!\,2!} \times \frac{3!}{1!\,2!} \times \frac{5!}{3!\,2!} =$$

$$\frac{8 \times 7}{2 \times 1} \times \frac{4 \times 3}{2 \times 1} \times \frac{3}{1} \times \frac{5 \times 4}{2 \times 1} = \frac{56}{2} \times \frac{12}{2} \times \frac{3}{1} \times \frac{20}{2} = 28 \times 6 \times 3 \times 10 = 5040$$

There are 5,040 different groups with 2 members of each class available to go to a conference.

Key words to look for to decide if it is a permutation or combination:

Permutation	**Combination**
arrangement	group
order	sample
schedule	committee

Practice Test Questions on Counting Principles, Permutations, and Combinations

1. How many ways could 5 desserts be sampled from a dessert buffet containing 9 different desserts?
   (a) 126
   (b) 45
   (c) 15,120
   (d) 120

2. A license plate in Pennsylvania consists of 3 letters followed by 4 numbers. If duplicates are allowed, how many different license plates are available?
   (a) 12
   (b) 260,000
   (c) 175,760,000
   (d) 24
3. A florist has decided to produce 6 different Valentine's Day flower arrangements. In how many different orders can she display these arrangements?
   (a) 36
   (b) 720
   (c) 1
   (d) 460
4. There are 3 chocolate cookies, 5 peanut butter cookies, and 4 shortbread cookies on a tray. Six cookies are selected at random to be put in a cookie bag. How many bags could contain 2 chocolate cookies, 2 peanut butter cookies, and 2 shortbread cookies?
   (a) 60
   (b) 665,280
   (c) 924
   (d) 180

Detailed answer key for the above problems:

1. Answer: (a) 126
   The order the desserts are selected does not matter. This is a combination (9,5).
2. Answer: (c) 175,760,000
   This problem is using the Fundamental Counting Principle. There are 26 letters available for each of the letters on the license plate and 10 numbers available for each of the numbers that will be on the license plate.

   $$26 \times 26 \times 26 \times 10 \times 10 \times 10 \times 10 = 175,760,000$$

3. Answer: (b) 720
   The order in which the flowers are arranged is important, so this is a permutation:

   $$P(6,6) = 720.$$

4. Answer: (d) 180
   The order the cookies are put in the bag is not important so this is a combination problem. The chocolate cookies are selected by taking 2 of the 3 cookies, the peanut butter cookies are selected by taking 2 of the 5 cookies, and the shortbread cookies are selected by taking 2 of the 4 cookies. These combinations are then multiplied together.

   $$C(3,2) \times C(5,2) \times C(4,2) = 180$$

Appendix A

Full Practice Test 1

Formulas you will need for the Math Test

Permutation: $_nP_r = \frac{n!}{(n-r)!}$

Combination: $_nC_r = \frac{n!}{(n-r)!r!}$

Simple Interest: $I = P \times r \times t$

Compound Interest: $A = P(1+r)^t$

Finding the midpoint: $\left(\frac{x_1+x_2}{2}, \frac{y_1+y_2}{2}\right)$

Calculating the distance between two points: $\sqrt{(x_2-x_1)^2 + (y_2-y_1)^2}$

Pythagorean Theorem: $c^2 = a^2 + b^2$

Rectangle: Area = lw Perimeter: $2l + 2w$

Triangle: Area = ½ bh

Circle: Area = πr^2 Circumference = $2\pi r$

Sphere:

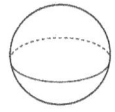

Surface Area = $4\pi r^2$
Volume = $\frac{4}{3}\pi r^3$

Cylinder:

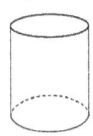

Surface Area = $2\pi rh + 2\pi r^2$
Volume = $\pi r^2 h$

Rectangular Solid:

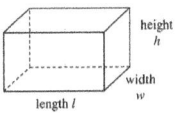

Surface Area = $2lw + 2lh + 2wh$
Volume = lwh

Appendix A

ANSWER SHEET FOR FULL PRACTICE TEST

Question	Answer Choice	Question	Answer Choice	Question	Answer Choice
1.		13.		25.	
2.		14.		26.	
3.		15.		27.	
4.		16.		28.	
5.		17.		29.	
6.		18.		30.	
7.		19.		31.	
8.		20.		32.	
9.		21.		33.	
10.		22.		34.	
11.		23.		35.	
12.		24.		36.	

The following table will divide the questions by chapter. Circle the questions you missed. This will help to identify your strengths and needs and will help focus study time where it is most needed.

Chapter 2 *Understanding Numbers and the Number System*	Chapter 3 *Pre-Algebra*	Chapter 4 *Algebra*	Chapter 5 *Measurement and Geometry*	Chapter 6 *Statistics and Probability*
QUESTIONS:	QUESTIONS:	QUESTIONS:	QUESTIONS:	QUESTIONS:
1	4	3	13	14
2	5	9	20	15
6	7	10	21	16
18	8	11	25	17
28	12	22	30	19
35	23		32	26
	24		36	
	27			
	29			
	31			
	33			
	34			

Appendix A

36 multiple choice questions

75 minutes maximum time

Directions: Read each item carefully and choose the best answer response.

1. Three out of four doctors recommend walking 30 minutes a day as the best solution to reducing your weight. What percent of doctors recommend something other than walking as the best solution to weight control?
   (a) 75%
   (b) 0.75
   (c) 0.25
   (d) 25%

2. What decimal below shows the number 7 in the millionths place?
   (a) 0.004667
   (b) 0.00476
   (c) 0.00746
   (d) 7,000,000

3. Which of the following equations listed below represents the given graph?

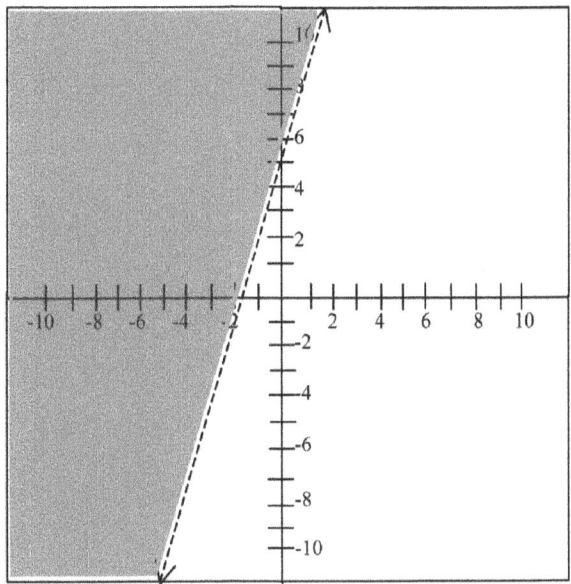

   (a) $y \geq 6x+10$
   (b) $2y \leq 6x-10$
   (c) $2y > 6x+10$
   (d) $y > x+3$

4. There are 18 students in a fourth-grade classroom. The teacher has planned a computer activity for her students to complete during the one-hour math class. There are 6 computers in the classroom for the students to share. Each student should

have equal time on a computer. How many minutes should each student have to complete the computer activity?
   (a) 60 minutes
   (b) 20 minutes
   (c) 40 minutes
   (d) 10 minutes
5. If it takes four days for ten roofers to lay a roof, how long would it take for eight roofers to complete the same job?
   (a) 3 days
   (b) 20 days
   (c) 5 days
   (d) 2 days
6. The price of one 16-ounce straightening conditioner went up from $20 to $40 in the past five years. What was the percent increase in the price of the conditioner in the past five years?
   (a) 100%
   (b) −50%
   (c) $20
   (d) 200%
7. Which of the following terms is a factor of $a^4b^2c + a^2bc^2$?
   (a) ab^2c
   (b) a^2bc
   (c) abc^2
   (d) a^4bc
8. Which of the following expressions is equal to $4a(3b^2-2)$?
   (a) $12ab^3-8$
   (b) $12ab^2-2$
   (c) $12ab^2-8a$
   (d) $4ab^2$
9. Which of the equations listed below best represents the graph that is given?

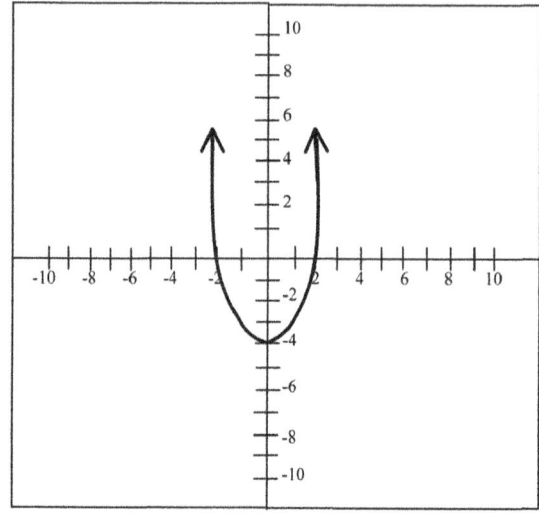

(a) $y=-x^2+4$
(b) $y=x-4$
(c) $y=x^2-4$
(d) $y=-x^2-4$

10. What is the equation of the line with the y-intercept 3 and the x-intercept 2?
    (a) $2x+3y=0$
    (b) $6x+4y=12$
    (c) $3x=2y$
    (d) $(2,3)$

11. Which of the following is the equation of a line that is parallel to the line described by the equation $y=2x+3$ and has a y-intercept of 4?
    (a) $4y=12x-4$
    (b) $12x+2y=0$
    (c) $(0,4)$
    (d) $-4x+2y=8$

12. A 450-meter fence is divided into three portions. The second portion is twice as long as the first. The third portion is three times as long as the second. Find the lengths of the portions.
    (a) 100 meters, 150 meters, 200 meters
    (b) 50 meters, 100 meters, 300 meters
    (c) 150 meters each
    (d) 100 meters, 200 meters, 300 meters

13. During spring break, the basketball team is traveling 72.5 kilometers for a tournament. They already traveled 35 kilometers. How many meters are left to travel?
    (a) 37,500
    (b) 37.5
    (c) 107.5
    (d) 35,000

14. The temperature of the Neonatal Intensive Care Unit (NICU) must average at least 72° for the unit to be considered a healthy environment. A mean temperature is taken every five days. The temperature on Monday was 74°, Tuesday 69°, Wednesday 73°, and Thursday 71°. What is the lowest temperature the NICU could be on Friday so that it considered healthy enough for the newborns?
    (a) 65°
    (b) 70.5°
    (c) 76°
    (d) 73°

15. The Student Leadership Council is selling T-shirts as an annual fundraising event. To simplify the ordering process, the council members created the T-shirt order form provided below. How many different T-shirt size, color, and design options are possible?

Appendix A

Student Leadership Council		
T-shirt Order Form NAME _____ CELL _____		
Size – check off choice	Color – check off choice	Design – check off choice
o SM o M o L o XL	o WHITE o BLUE o GOLD	o LOGO o MOTTO

    (a) 50
    (b) 9
    (c) 24
    (d) 3

16. An integer from 1 to 40 is generated at random. What is the probability the number generated is less than 10 or greater than or equal to 30?
    (a) 0.5
    (b) 0.2
    (c) 0.95
    (d) 20

17. In how many different ways can the letters T, E, A, C, H, I, N, G be arranged?
    (a) 8
    (b) 40,320
    (c) 36
    (d) 100,000

18. Find a counterexample for the following statement: If a number is divisible by 3, then it is also divisible by 6.
    (a) 18
    (b) 16
    (c) 15
    (d) 6

19. In a survey of 120 college juniors, the following information was collected: 30 juniors take Physics, 40 juniors take Biology, and 53 take Calculus I. Six juniors take Physics and Biology, 13 take Biology and Calculus I, and 15 take Physics and Calculus I, while 5 juniors take all three subjects. How many college juniors do not take any of the three subjects? (Hint: Construct a Venn Diagram to help answer the question.)
    (a) 3
    (b) 94
    (c) 26
    (d) 130

20. The base of a 10-foot ladder is placed 6 feet from a wall. The top of the ladder is leaning against the wall of the building. How far up the wall does the ladder reach?
    (a) eight feet
    (b) ten feet
    (c) twenty-five feet
    (d) twelve feet

21. Given the parallelogram below, what is the measure of Angle CDE?

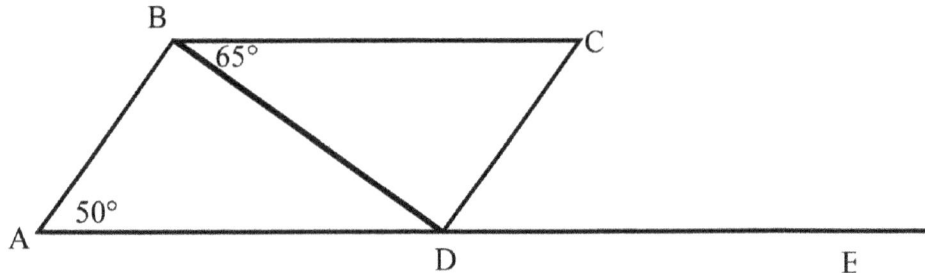

(a) 50°
(b) 150°
(c) 90°
(d) 20°

22. What is the solution to the following system of equations?

$$y-x=5$$
$$y=2x+1$$

(a) (5,1)
(b) (1, –2)
(c) $x=4$
(d) (4,9)

23. Identify the pattern in the sequence below. If this pattern continues, how many tally marks will make up figure 6 in the sequence?

Figure 1   Figure 2   Figure 3   Figure 4

(a) 32
(b) 64
(c) 128
(d) 16

24. The All-City Cab Company charges a flat rate of $3.00 plus $1.40 per mile. Which of the following equations gives the total cost in dollars, y, for a ride of x miles?
(a) $x=3y+1.4$
(b) $y=1.4x+3$
(c) $y=x+1.4$
(d) $y=3x+1.4$

25. A circle is drawn on the coordinate plane. The center of the circle is situated at (–4,–5) and the edge of the circle passes through (–4,3). What is the area of the circle?
(a) 16π
(b) 8π
(c) 64π
(d) 4π

26. Describe the type of graph that would best illustrate stock market data for a month.
    (a) bar graph
    (b) line graph
    (c) circle graph
    (d) box-and-whisker plot
27. If the sum of two numbers is 550 and their ratio is 10:1, what is the smaller number?
    (a) 50
    (b) 500
    (c) 600
    (d) 5
28. Using the number line below, which of the following expressions results in the least value?

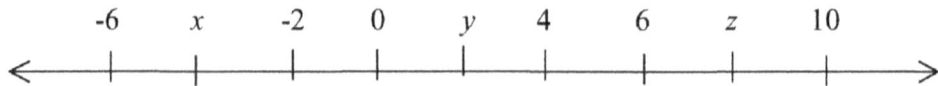

    (a) $x + y + z$
    (b) $10xyz$
    (c) $10x + 3y$
    (d) $-10(x + y + 5z)$
29. If $300 is divided among three little league teams in the ratio of 2:3:7, what is the difference between the least amount and the greatest amount divided among the teams?
    (a) $50
    (b) $25
    (c) $125
    (d) $175
30. If Quadrilateral ABCD below is reflected over the x-axis, what are the coordinates of the image point B'?

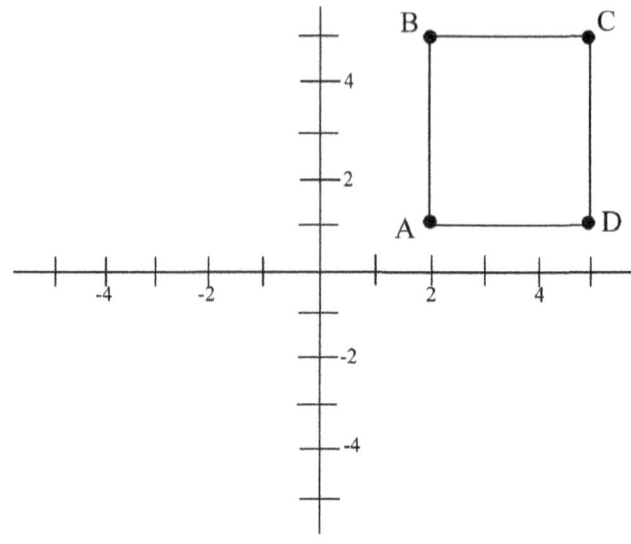

(a) (2,−1)
(b) (−2,5)
(c) (2,5)
(d) (2,−5)

31. The ratio of commercial time to total airtime of a weekly sitcom is 3:15. If a weekly sitcom airs for 30 minutes, which of the following proportions could be used to calculate the number of commercial minutes, x, during the 30-minute timeslot?
    (a) $\frac{3}{30}=\frac{15}{x}$
    (b) $3x = 30$
    (c) $\frac{3}{15}=\frac{30}{x}$
    (d) $\frac{3}{15}=\frac{x}{30}$

32. A gardener has a square piece of land that measures 15 feet on a side. He wants to divide the land into 25 equal squares. How big will each square be?
    (a) 3 feet × 3 feet
    (b) 5 feet × 5 feet
    (c) 3 feet × 5 feet
    (d) 10 feet × 10 feet

33. Drew is three years older than Gloria, and Chris is three years older than Drew. The sum of their ages is 78. How old is Drew?
    (a) 23
    (b) 26
    (c) 29
    (d) 40

34. Continue the pattern in the table below. Identify the value of the sixth term:

Term	Value of the Term
1	2
2	8
3	24
4	64
5	160
6	—

   (a) 100
   (b) 896
   (c) 384
   (d) 900

35. Some values of x are less than 60. All of the statements below are consistent with this statement except:
    (a) no numbers less than 60 are values of x
    (b) 65 can be a value of x
    (c) no numbers greater than 60 are values of x
    (d) all values of x are less than 60

36. Use the diagram below to answer the question that follows:

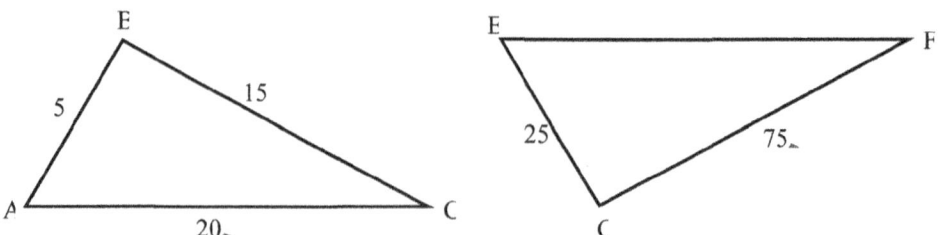

In the drawing above, triangle ABC is similar to triangle EGF. Which of the following proportions will calculate the length of side EF?

(a) $\dfrac{15}{5} = \dfrac{EF}{25}$

(b) $\dfrac{EF}{5} = \dfrac{75}{15}$

(c) $\dfrac{5}{25} = \dfrac{20}{EF}$

(d) $\dfrac{5}{20} = \dfrac{EF}{25}$

ANSWERS AND DETAILED EXPLANATIONS
FOR FULL PRACTICE TEST 1

1. Answer: (d) 25%

   Three out of four doctors recommend walking. Three out of four = $\tfrac{3}{4}$, which is equivalent to 75%. The remaining doctors recommend something other than walking as a solution to weight control. Since percent means out of a hundred, then simply subtract 100−75=25. The remaining percentage represents the doctors who do not recommend walking as the best solution. Choice (c) gives the answer in decimal form. Choice (a) is the percent of those who recommend walking, and choice (b) gives that answer as a decimal. Choice (d) is the only option that answers the question.

   Topic of Question 1: Applying percents. More on percents can be found in chapter 2.

2. (a) 0.004667

   The place values to the right of the decimal point are as follows: *tenths, hundredths, thousandths, ten-thousandths, hundred-thousandths, millionths, etc.* The millionths place is the sixth place to the right of the decimal point. Choice (a) is the only choice with a 7 in the sixth place to the right of the decimal point (millionths place). Answer choice (d) is incorrect because it shows 7 million, not 7 millionths, and can be automatically eliminated. Answer choice (b) has a seven in the ten-thousandths place, and answer choice (c) has a seven in the thousandths place. Both (b) and (c) can also be eliminated.

   Topic of Question 2: Understanding place value. More on place value can be found in chapter 2.

3. Answer: (c) 2*y*>6*x*+10

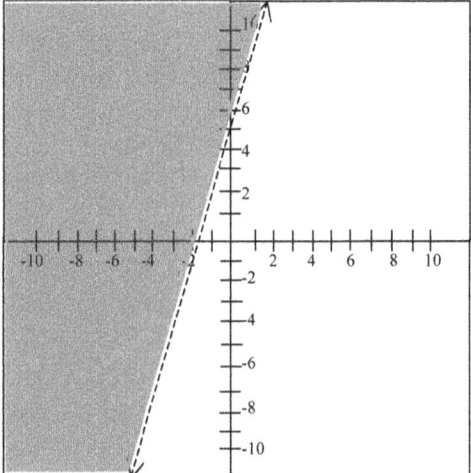

First, it is important to notice the dashed line in the graph instead of a solid line. That indicates all the points on the line itself are not included as part of the solution set. Thus, options (a) and (b) can be eliminated. Next, convert option (c) to slope-intercept form by solving in terms of y.

$$2y > 6x + 10$$
$$y > 3x + 5$$

Recognize that the y-intercept in both answer choice (c) and the graph is 5. Use the slope $\frac{3}{1}$ in the inequality to check for accuracy. Start at the y-intercept $(0,5)$ and move one point to the right and three points up to discover that point $(1,8)$ is a solution to the inequality. Next, substitute an easy point to discover if the correct half-plane represented by the inequality is the same as the one determined in the graph. Substitute point $(-10,0)$ to see if it makes the inequality true:

$$y > 3x + 5$$
$$0 > 3(-10) + 5$$
$$0 > -25$$

Answer choice (c) $2y > 6x + 10$ is correct.

Topic of Question 3: Graphing inequalities. More on inequalities be found in chapter 4.

4. Answer: (b) 20 minutes

   Convert 1 hour to 60 minutes. Multiply the total number of minutes by the number of available computers: 60 minutes × 6 computers = 360 total minutes available for computer use. Divide 360 by the number of students to calculate the number of available computer minutes per student: 360 total minutes ÷ 18 students in the class = 20 minutes per student. Answer choice (a) would only make sense if there were 18 computers since 60 minutes is the total time available. This answer choice can be eliminated as a possible option. Answer choice (d) does not make sense either since only providing each student with 10 minutes of computer time would result in another 180 minutes of spare time with no students on computers. This answer can also be eliminated.

Topic of Question 4: Applying algebraic principles in word problems. More on algebraic principles in word problems can be found in chapter 3.

5. Answer: (c) 5 days

    Answer choices (a) and (d) can be automatically eliminated because it does not make sense for fewer roofers to complete a job in fewer days. Answer choice (b) is also not a reasonable answer because it would not take 20 days to do a job with only two fewer workers than doing the job in 4 days. Answer choice (b) is the only real reasonable answer. Now do the calculations to make sure it is the correct answer.

    This is an indirect proportion, not a direct proportion, problem because the fewer people who work on a job, the longer it will take to complete the job.

    In an indirect proportion, make an equation in the following way:

    $$10 \text{ roofers} \times 4 \text{ days} = 8 \text{ roofers} \times d \text{ days}$$
    $$40 = 8d$$

    Divide by 8 to solve for d. It would take eight roofers 5 days to complete the same job.

    None of the other answer choices make sense.

    Topic of Question 5: Applying algebraic principles in word problems. More on algebraic principles in word problems can be found in chapter 3.

6. Answer: (a) 100%

    First, estimate the increase. The cost for one bottle was $20, and in five years, it was $40. That means five years ago two bottles would cost the same as one bottle five years later. That is a 100% increase because it cost the same for one bottle that it originally cost for two. Now, start the calculations to see if the estimate is correct. To find percent increase, first calculate the amount of increase. $40−$20=$20. Now divide the amount of increase by the original amount. 20÷20=1. Convert to a percent by multiplying 1 by 100. 1 × 100=100%. The percent of increase is 100%. Go back to the estimate to see if the answer makes sense. Yes, it makes sense that something that was originally $20 and is now $40 is a 100% increase.

    Choice (b) doesn't make sense because the cost did not decrease. Choice (c) is simply the amount of increase and not the percentage of increase. Choice (d) is too high of an increase.

    Topic of Question 6: Applying percent of increase. More on percents can be found in chapter 2.

7. Answer: (b) a^2bc

    a^2 is a factor of both a^4 and a^2. b is a factor of both b^2 and b. c is a factor of both c and c^2. Answer choice (a) doesn't work because c^2 is not a factor of the second term. Answer choice (c) is incorrect because c^2 is not a factor of the first term. Answer choice (d) is incorrect because a^4 is not a factor of the second term.

    Topic of Question 7: Factoring. More on factoring can be found in chapter 3.

Appendix A

8. Answer: (c) $12ab^2-8a$

   The terms in the parentheses are not like terms so they cannot be combined. Use the distributive property and multiply each term in the parentheses by $4a$. When multiplying like terms, add the exponents.

   $$4a \times 3b^2 = 12ab^2$$
   $$4a \times 2 = 8a$$

   The expression equal to $4a(3b^2-2)$ is $12ab^2-8a$. Answer choice (a) is wrong because the first term distributed should be $12ab^2$ and not $12ab^3$. Answer choice (b) is wrong because $4a$ must be multiplied by both terms in the parentheses and this choice did not multiply $4a$ and 2 to get $8a$. Answer choice (d) is wrong because the two terms in the parentheses cannot be combined as they are unlike terms.

   Topic of Question 8: Using the distributive property. More on distribution can be found in chapter 3.

9. Answer: (c) $y=x^2-4$

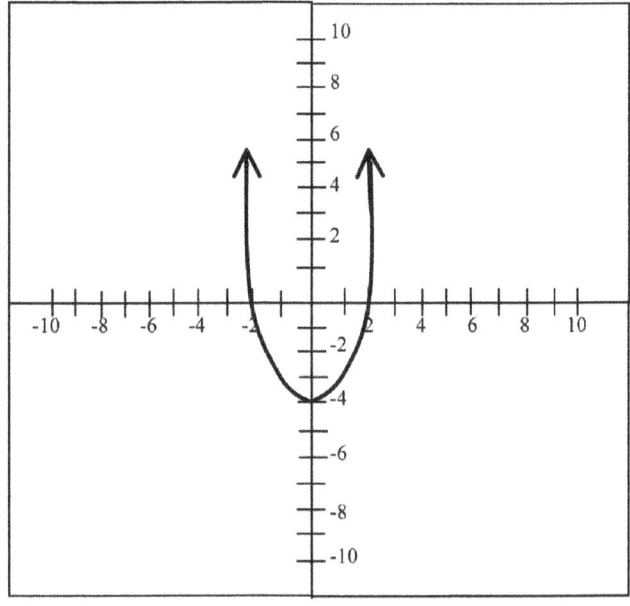

   Because the graph includes a vertical parabola, the value of x must be squared. Therefore, answer choice (b) is eliminated. It is also true that because the parabola opens upward, the coefficient of the x^2 term must be positive, so that eliminates both answer choices (a) and (d). The vertex of the parabola shows the y-intercept, which is at -4. The only answer choice that is reasonable is choice (c) $y=x^2-4$. Now, substitute $(0,-4)$ into the equation to check for accuracy.

   $$-4=-(0)^2-4$$
   $$-4=-4$$

Answer choice (c) $y=x^2 - 4$ is correct.
Topic of Question 9: Graphing parabolas. More on parabolas can be found in chapter 4.

10. Answer: (b) $6x+4y=12$

    Using the intercept for $x=(2,0)$ and $y=(0,3)$, substitute both points to solve each equation to determine which answer choice is true for both intercepts.
    Answer choice (d) represents a point and can be eliminated because it is not an equation.
    Answer choice (a) is not a true equation when the intercepts are substituted.

$$2x+3y=0$$

Substitute the *x*-intercept (2, 0)

$$2(2)+3(0)\neq 0$$

It is not the equation of the line with the *x*-intercept of 2.
Answer choice (c) also does not work when substituting the *x* and *y* intercepts.
Answer choice (b) is the equation of the line with the *y*-intercept of 3 and the *x*-intercept of 2. This can be proven through substitution of both intercepts.

$$6x+4y=12$$
$$6(2)+4(0)=12$$
$$12=12$$

This is a true equation using the *x*-intercept.

$$6(0) + 4(3) =12$$
$$12=12$$

This is a true equation using the *y*-intercept.
To check the answer, locate the *x*-intercept. Set *y* at 0 and solve the equation for *x*.

$$6x+4(0) =12$$
$$6x=12$$
$$x=2$$

The line passes through the *x*-axis at (2,0).
To locate the *y*-intercept, set *x* at 0 and solve the equation for *y*.

$$6(0) + 4y=12$$
$$4y=12$$
$$y=3$$

The line passes through the *y*-axis at (0,3).
Topic of Question 10: x- and y-intercepts. More on this can be found in chapter 4.

11. Answer: (d) $-4x+2y=8$
    A line in the form $y=mx+b$ is in slope-intercept form where m = the slope and b is the y-intercept. The slope of the line $y=2x+3$ is 2. The slope of the equation $y=2x+4$ is 2. Lines that are parallel to one another have the same slope.
    Answer choice (c) does not answer the question because it is a point and not an equation of a line. This answer can be automatically eliminated.
    Next, solve the remaining answer choices in terms of y in order to convert the equations to slope-intercept form.
    Answer choice (a) in slope-intercept form is $y=3x-1$. The slope is 3. It is not parallel to a line with the equation $y=2x+3$.
    Answer choice (b) in slope-intercept form is $y=-6x$.
    This equation has a slope of -6 and does not have a y-intercept. This cannot be the right answer either.
    Answer choice (d) is the only option where the slope is the same as given line and has a y-intercept of 4
    Convert $-4x+2y=8$ to slope-intercept form:

$-4x+2y=8$	Solve for y by adding $4x$ to each side of the equation.
$2y=4x+8$	Divide each term by 2.
$y=2x+4$	2 is the slope, which is parallel to the given equation $y=2x+3$.

    The y-intercept of $y=2x+4$ is 4.
    Answer choices (a) and (b) are not parallel to the given line and do not have y-intercepts of 4.
    Topic of Question 11: Graphing parallel lines. More on this can be found in chapter 4.

12. Answer: (b) 50 meters, 100 meters, 300 meters
    Create an algebraic equation to calculate the length for each portion of fence. Since the information is given in terms of the first portion of fence, x will represent the length of the first portion. The second portion is twice as long as the first, which would be represented as $2x$. The third portion is three times as long as the second: $3(2x)$ or $6x$. The equation is as follows:

$x+2x+3(2x)=450$	Collect like terms on the left side of the equation.
$9x=450$	Divide by 9.

    $x=50$ or the length of the smallest portion of the fence.
    Replace 50 for x in the equation and calculate the lengths for the remaining two portions of fence.
    $x = 50$ meters, $2x = 100$ meters, and $6x = 300$ meters.
    Answer choices (a) and (c) add up to 450 meters but are incorrect lengths of each portion. The second length of each (a) and (c) is not twice as long as the first and the third length is not three times the length of the second length. Both answer choices (a) and (c) can be eliminated. Answer choice (d) does not add up to 450 and can also be eliminated.
    Topic of Question 12: Applying algebraic principles in word problems. More on algebraic principles in word problems can be found in chapter 3.

13. Answer: (a) 37,500

    The basketball team will be traveling 72.5 kilometers in all. They already traveled 35 kilometers. They still need to travel 72.5 − 35 = 37.5 kilometers. Convert kilometers to meters by multiplying the answer by 1,000: 37.5 × 1000 = 37,500 meters.

    The basketball team needs to travel 37,500 more meters to get to the tournament. Answer choice (b) gives the answer in kilometers, which doesn't answer the question. Answer choice (c) gives the total kilometers of the trip added to the kilometers already traveled. This also doesn't answer the question. Answer choice (d) represents the 35 kilometers already traveled in meters.

    Topic of Question 13: Understanding the metric system of measurement. More on using the metric system can be found in chapter 5.

14. Answer: (d) 73°

    Estimate a reasonable fifth temperature. The temperature on Monday was 74°, Tuesday 69°, Wednesday 73°, and Thursday 71°. In order to keep a mean temperature of 72°, the fifth temperature would have to be somewhere between 72 and 74. This makes sense because the temperatures are close to the expected mean and the range is only 4°.

    Create an equation to calculate the fifth score: $\frac{74+69+73+71+x}{5} = 72$

    Collect like terms and multiply both sides by 5 to clear the fraction.

    $$287 + x = 360$$

    Subtract both sides of the equation by 287.

    $$x = 73$$

    Another strategy is to use each of the answers provided to calculate what the mean would be with the fifth number included in the data: $\frac{74+69+73+71+73}{5} = 72$.

    For the mean to be 72°, answer choice (a) wouldn't make sense because it is too low for the average of the five temperatures to be 72° and (c) isn't reasonable because it is too high for the average to be 72°. Answer choice (b) doesn't make sense either because it is the mean of the four days without a fifth day included.

    Topic of Question 14: Applying measures of central tendency. More on measures of central tendency can be found in chapter 6.

15. Answer: (c) 24

    The Fundamental Counting Principle states that if there are *m* ways for one event to occur, and *n* ways for a second event to occur, then there are *m* × *n* ways for both to occur. Since there are four different size options, three different colors, and two possible designs, then according to the Fundamental Counting Principle, there are 4 × 3 × 2 = 24 possible T-shirt combinations. Answer choices (b) and (d) do not make sense as possible options because there can't be fewer combinations than there are possible options.

    Topic of Question 15: Applying the fundamental counting principle. More on the Fundamental Counting Principle can be found in chapter 6.

16. Answer: (a) 0.5

    There are 40 possible outcomes when generating an integer from 1 to 40. Nine of the forty are integers less than 10 and eleven of the 40 are greater than or equal to 30. When finding the probability of one event *or* another, add the probabilities of each: $\frac{9}{40} + \frac{11}{40} = \frac{20}{40} = \frac{1}{2} = 0.5$.

    Answer choice (d) is not a probability. A probability must be a number between 0 and 1, inclusive. Answer choice (d) 20 is the total number in the sample size. It is not written as a probability of the occurrence over the total number possible. Answer (b) is incorrect because the probability of generating a number equal to or greater than 30 alone is greater than 0.2. Answer (c) doesn't make sense as an answer because it is unreasonable to think that the probability described would be as big as 0.95 and can be eliminated.

    Topic of Question 16: Probability. More on probability can be found in chapter 6.

17. Answer: (b) 40,320

    There are 8! different ways to arrange the letters T, E, A, C, H, I, N, G.
    8! = 8 × 7 × 6 × 5 × 4 × 3 × 2 × 1 = 40,320
    No other answer choice is a reasonable answer to this problem.

    Topic of Question 17: Factorials. More on factorials can be found in chapter 6.

18. Answer: (c) 15

    A counterexample is a case for which a statement is *not* true. Try each answer choice. Answer choices (a), (b), and (d) are all divisible by both 3 and 6 and therefore are not counterexamples to the statement. Answer choice (c), although divisible by 3, is not divisible by 6. Fifteen is a counterexample of the statement: If a number is divisible by 3, then it is also divisible by 6.

    Topic of Question 18: Number relationships. More on number relationships can be found in chapter 2.

19. Answer: (c) 26

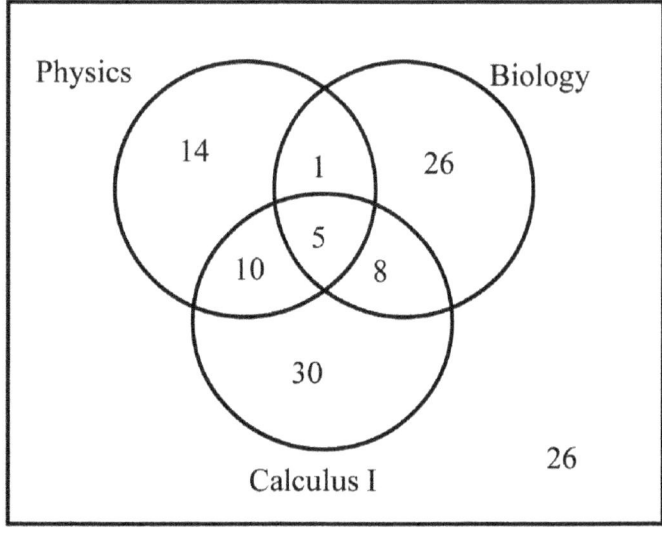

Construct a Venn Diagram describing the results of the survey. Always begin with the innermost section of the Venn Diagram—in this case the five juniors that take all three subjects. Then work out and complete the sections where juniors take only two of the three subjects by subtracting 5 from those numbers (6 – 5 = 1 took both Physics and Biology, 15 – 5 = 10 took Physics and Calculus I, and 13 – 5 = 8 took Biology and Calculus I). In order to calculate the juniors that took only one subject, add each section within the subject circles filled in so far and subtract that from the number of juniors taking each subject (30 – 16 = 14 juniors took Physics only, 40 – 14 = 26 juniors took Biology alone, and 53 – 23 = 30 juniors took Calculus I alone). Add all of the sections and subtract that from the total number of juniors surveyed to find the number of juniors who did not take any of the subjects (120 – 94 = 26 students did not take any of the three subjects). *Topic of Question 19: Sets and Venn Diagrams. More on sets and Venn Diagrams can be found in chapter 6.*

20. Answer: (a) eight feet

The 10-foot ladder, the ground, and the wall form a right triangle. Use the Pythagorean Theorem to solve the problem: In any right triangle, the sum of the squares of the lengths of the legs is equal to the square of the length of the hypotenuse. First, estimate your answer. Answer choices (b), (c), and (d) do not make sense because the opposite side of the right angle (or hypotenuse) of a right triangle is the longest side of the triangle. Answer choice (a) is the only answer that really makes sense. Do the calculations to check your estimate.

$$c^2 = a^2 + b^2$$
$$10^2 = 6^2 + b^2$$
$$100 = 36 + b^2$$

Subtract 36 from both sides of the equation.

$$b^2 = 64$$

Take the square root of 64.
$b = 8$ feet

Topic of Question 20: The Pythagorean Theorem. More on estimating and calculating measurements using the Pythagorean Theorem can be found in chapter 5.

21. Answer: (a) 50°

First, estimate the measure of angle CDE. It is clearly an acute angle, which eliminates answer choices (b) and (c). Knowing that the sum of the angles of a parallelogram are 360° and adjacent angles of a quadrilateral are supplementary, the measure of angle ABD would be 65° since the sum of angles ABD and DBC must total 130°.

The sum of the angles of a triangle is 180°, so the measure of angle ADB is 65° since angle BAD = 50° and angle ABD = 65°. Therefore, the measure of angle BDC is also 65°. Angles ADC and CDE form a linear pair; angle ADC = 130°. Therefore, angle CDE = 50°.

Topic of Question 21: Attributes of triangles and angle relationships. More on triangles can be found in chapter 5.

Appendix A

22. Answer: (d) (4,9)
    One strategy is to first rename each equation in standard form: $Ax + By = C$

    $$y - x = 5$$
    $$y - 2x = 1$$

    The goal is to eliminate one of the variables. In order to eliminate the x, multiply each term in the first equation by -2 and add the equations. This will eliminate the x variable.

    $$-2(y - x) = -2(5)$$

    Multiplying each term by the same value does not change the equation. Now add the two equations.

    $$\begin{array}{r} -2y + 2x = -10 \\ +y - 2x = 1 \\ \hline -y = -9 \\ y = 9 \end{array}$$

    Substitute 9 in for y in the top equation and solve for x: $9 - x = 5$.

    $$x = 4$$

    The solution is (4,9).
    Another strategy is to use the answers in the given equations and find the ordered pair that makes both equations true.
    Substitute the values for x and y from answer choice (a) into the first equation:
    (a) (5,1) $1 - 5 \neq 5$
    Substitute the values for x and y from answer choice (b) into the first equation:
    (b) (1, –2)

    $$-2 - 1 \neq 5$$

    Substitute the values for x and y from answer choice (d) into the first equation:
    (d) (4,9) $9 - 4 = 5$
    Then check the second equation by replacing $x = 4$ and $y = 9$ in both equations to check your answer.

    $$9 - (2 \times 4) = 1$$
    $$9 - 8 = 1$$

    Answer choice (c) is not a point and can be eliminated. Answer choices (a) and (b) do not make either equation true.
    Topic of Question 22: Solving systems of equations. More on systems of equations be found in chapter 4.

23. Answer: (b) 64
    First identify the pattern sequence: n = figure number 2^n = pattern sequence.

Therefore, the following sequence would continue:

Figure 1 = 2^1 = 2 tally marks
Figure 2 = 2^2 = 4 tally marks
Figure 3 = 2^3 = 8 tally marks
Figure 4 = 2^4 = 16 tally marks
Figure 5 = 2^5 = 32 tally marks
Figure 6 = 2^6 = 64 tally marks

Answer choice (a) 32 is the number of tally marks for the fifth figure. Answer (c) 128 represents the seventh figure, and answer (d) 16 is already shown as the number of tally marks in the fourth figure.

Topic of Question 23: Applying algebraic principles and function tables. More on function tables can be found in chapter 3.

24. Answer: (b) y = 1.4x + 3

The equation must be set to the total cost in dollars, y, for a ride of x miles. Answer choice (a) can be immediately eliminated because it is not a solution in terms of y, the total cost. The fixed cost is the $3.00 flat fee and is not affected by the number of miles, which is represented by the variable x. $1.40 is the marginal cost, because the cost changes according to the number of miles traveled, so the cost is dependent upon the number of miles (1.4x).

Answer choice (c) includes the marginal cost as the fixed cost and does not include a cost per mile at all. Answer choice (d) is incorrect because the fixed and marginal costs are swapped in the equation.

Topic of Question 24: Applying algebraic principles in word problems. More on interpreting algebraic principles from word problems can be found in chapter 3.

25. Answer: (c) 64π

Find the area of a circle whose center is situated at (–4,–5) and edge passes through (–4,3). The formula for area of a circle is $A=\pi r^2$.

Find the difference between the two *y*-values to determine the radius. The difference between –5 and 3 is 8 units. It is helpful to draw a quick quadrant grid and graph the two points. This will help distinguish that the radius is 8 units.

Therefore, the area is

$$8^2\pi = 64\pi$$

Answer choice (a) is incorrect because $4^2=16$, which is not the full length of the radius. Four units will only reach to the *x*-axis, not the edge of the circle. Answer choice (b) 8π identifies the length of the radius but does not follow the formula of the area of a circle because 8 would have to be squared first. Answer choice (d) does not make any sense because it is not the correct length of the radius and it is not accurately simplified in the formula for the area of a circle. Therefore, it can be automatically eliminated.

Topic of Question 25: Area and attributes of a circle. More on circles can be found in chapter 5.

26. Answer: (b) line graph
A line graph is a graph that shows changes in data over time. This would be the most appropriate graph to display stock market data over a month. Answer choice (a) is incorrect because a bar graph is best used to show relationships or comparisons between categories. Answer choice (c) is wrong because a circle graph is best used to compare parts to the whole and parts to other parts. Answer choice (d) is incorrect because a box-and-whisker plot shows the distribution and spread of data.
Topic of Question 26: Choosing appropriate ways to show data. More on appropriateness of graphs can be found in chapter 6.

27. Answer: (a) 50
Create an equation to find the smaller number. The sum of two numbers is 550 and their ratio is 10:1.
x = the smaller number and $10x$ = the larger number. Therefore, $x + 10x = 550$. Collect like terms: $11x = 550$. Then divide by 11: $x = 50$.
Replace the smaller number back into the equation to see if the answer is correct: $50 + 10(50) = 550$. It is correct. The value of the smaller number is 50 and the value of the larger number is 500. The ratio 500:50 is a 10:1 ratio and the sum of 500 and 50 is 550. Choice (a) 50 is correct.
Answer choices (b) and (c) are not reasonable answers because the smaller number cannot be either 500 or 600 since both would result in a total greater than 550. Both answer choices can be automatically eliminated. Answer choice (d) also doesn't make sense in the problem because $5 + 10(5) = 55$, not 550.
Topic of Question 27: Applying algebraic principles in word problems. More on algebraic principles in word problems can be found in chapter 3.

28. Answer (b) $10xyz$
First identify the interval (space between the tick marks). The interval is 2.
Find the value of each of the variables: $x = -4$, $y = 2$, and $z = 8$.
Substitute the values of each of the variables and evaluate each expression
(a) $x + y + z = -4 + 2 + 8 = 6$
(b) $10xyz = 10 \times (-4) \times 2 \times 8 = -640$
(c) $10x + 3y = 10(-4) + 3(2) = -40 + 6 = -34$
(d) $-10(x + y + 5z) = -10(-4 + 2 + 5 \times 8) = 10(-38) = -380$
Answer choice (b) $10xyz = 10 \times (-4) \times 2 \times 8 = -640$ is the least value.
Topic of Question 28: Relative magnitude of real numbers. More on real numbers can be found in chapter 2.

29. Answer: (c) $125
$300 is divided among three little league teams in the ratio of 2:3:7. That means one team will receive 2 parts, another 3 parts, and the third team will receive 7 parts of the total amount of money dispersed to the three teams. Create an equation in order to calculate the value of one part as described in the problem.

$$2x + 3x + 7x = 300$$

Collect like terms: $12x = 300$
Divide by 12.

$$x = 25$$

One part of the money is equal to $25.
Substitute the value of each part back in the equation.

$$2(25) + 3(25) + 7(25) = 300$$

Subtract the least amount from the greatest amount.

$$\text{Greatest amount: } 7 \times 25 = 175$$
$$\text{Least amount: } \quad 2 \times 25 = 50$$
$$175 - 50 = \$125$$

$125 is the difference between the least and greatest amount of money divided among the teams.

Answer choice (a) is incorrect. It represents the least amount of money one of the teams will receive. It does not answer the question. Answer choice (b) also does not answer the question. It is equal to what each part represents. Answer choice (d) is wrong because it describes the greatest amount ($7x$) one of the teams will receive and not the difference. It also does not answer the question. Answer choice (c) is the only reasonable answer.

Topic of Question 29: Applying algebraic principles in word problems. More on algebraic principles in word problems can be found in chapter 3.

30. Answer: (d) (2,–5)

    Quadrilateral ABCD reflected over the x-axis would result in an image with the same x-coordinate for each image point and an opposite y-coordinate for each image point. Therefore, the image of point B reflected over the x-axis would be (2,–5). Answer choice (a) (2,–1) is the image of point A reflected over the x-axis. It can be eliminated as a possible answer. Answer choice (b) (–2,5) would be the coordinates of B' if it was reflected over the y-axis. Answer choice (c) (2,5) is the coordinates for point B before the reflection. The only answer choice that is reasonable is (d) (2,–5).

    Topic of Question 30: Graphing transformations. More on transformations can be found in chapter 5.

31. Answer: (d) $\frac{3}{15} = \frac{x}{30}$

    A proportion is an equation showing that two ratios or rates are equal. The ratio of commercial time to total airtime of a weekly sitcom is 3:15. The ratio comparison will be commercials over the total airtime for the sitcom. If a weekly sitcom airs for thirty minutes, which indicates total airtime, the number of commercial minutes, x, would represent the numerator of the second ratio.

    Answer choices (a) and (c) do not form proportions that include the equivalent of 3:15 commercials to airtime. Answer choice (b) is not a proportion and can be automatically eliminated.

    Topic of Question 31: Applying algebraic principles in word problems involving proportions. More on algebraic principles in word problems can be found in chapter 3.

32. Answer: (a) 3 feet × 3 feet

    A gardener has a square piece of land that measures 15 feet on a side. The area of the total garden is 225 feet^2. He wants to divide the land into 25 equal squares.

Divide 225 by 25, which is the area of each square: 9 ft². Since 9 is the area of each square, find the square root of 9. The square root of 9 is 3. Each of the 25 squares will be 3 feet × 3 feet. Another method is to find the square root of 25 and divide that into the measurement of one of the sides: square root of 25 = 5 and 15 ÷ 5 = 3. Each side will be 3 feet long.

Answer choice (c) is not a square and can be automatically eliminated. Answer choices (b) and (d) do not answer the question and can also be eliminated as possible answers.

Topic of Question 32: Understanding the relationship between area and perimeter. More on relating area and perimeter can be found in chapter 5.

33. Answer: (b) 26

Drew is three years older than Gloria and Chris is three years older than Drew. The sum of their ages is 78. Create an equation representing this information in order to find the age of Gloria, the youngest of the three people.

$$x = \text{Gloria's age}$$
$$x + 3 = \text{Drew's age}$$
$$x + 3 + 3 = \text{Chris's age}$$
$$(x) + (x + 3) + (x + 3 + 3) = 78$$

Collect like terms: $3x + 9 = 78$.

Subtract 9 from both sides of the equation. $3x = 69$. Divide by 3 to get $x = 23$.

Gloria is 23 years old, which means Drew is $23 + 3 = 26$ years old.

Answer choice (a) is Gloria's age and answer choice (c) is Chris's age. Both choices do not answer the question. Answer choice (d) is not a reasonable answer and can be automatically eliminated.

Topic of Question 33: Applying algebraic principles in word problems. More on algebraic principles in word problems can be found in chapter 3.

34. Answer: (c) 384

In order to identify the sixth value of the term, first identify the geometric pattern in the table below:

Term	Value of the Term
1	2
2	8
3	24
4	64
5	160
6	—

Term	Value of the Term
1	2
2	8
3	24
4	64
5	160
6	384

The pattern is to take each term number and multiply it by $2^{\text{term number}}$.

$$1 \times 2^1 = 2$$
$$2 \times 2^2 = 8$$
$$3 \times 2^3 = 24$$
$$4 \times 2^4 = 64$$
$$5 \times 2^5 = 160$$
$$6 \times 2^6 = 384$$

Answer choice (a) can be automatically eliminated because the value is smaller than the value of the fifth term. Answer choice (b) is the value of the seventh term. It can also be eliminated because it does not answer the question. Answer choice (d) is not a reasonable answer because it does not follow the pattern or even resemble the pattern.

Topic of Question 34: Identifying patterns. More on patterns can be found in chapter 3.

35. Answer: (a) no numbers less than 60 are values of x

Answer choice (a) no numbers less than 60 are values of x is the only option that contradicts the statement "Some values of x are less than 60."

If some values of x are less than 60, then 65 can be a value of x.

Some values of x are less than 60 does not state that any numbers have to be greater than 60, so answer choice (c) still makes the statement true.

Answer choice (d) all values of x are less than 60 is still true if some values of x are less than 60.

Topic of Question 35: Understanding numbers and the number system. More on numbers and the number system can be found in chapter 2.

36. Answer: (c) $\frac{5}{25} = \frac{20}{EF}$

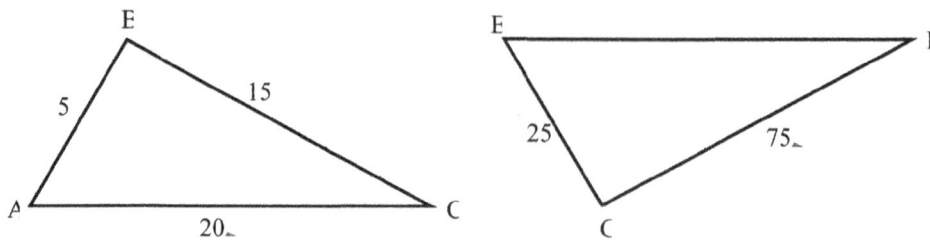

Since triangle ABC and triangle EGF are similar, their corresponding sides form proportions. It is important to first match up the corresponding sides for the triangles. Side AB corresponds to side EG. Side BC corresponds to side GF. Side AC corresponds to side EF. Therefore the following proportions are correct:

$$\frac{5}{25} = \frac{20}{EF} = \frac{15}{75}$$

Answer choice (c) is the only option where the ratios are equal.

Appendix B

Full Practice Test 2

Formulas you will need for the Math Test

Permutation: $_nP_r = \frac{n!}{(n-r)!}$

Combination: $_nC_r = \frac{n!}{(n-r)!r!}$

Simple Interest: $I = P \times r \times t$

Compound Interest: $A = P(1 + r)^t$

Finding the midpoint: $\left(\frac{x_1+x_2}{2}, \frac{y_1+y_2}{2}\right)$

Calculating the distance between two points: $\sqrt{(x_2 - x_1)^2 + (y_2 - y_1)^2}$

Pythagorean Theorem: $c^2 = a^2 + b^2$

Rectangle: Area = lw Perimeter: $2l + 2w$

Triangle: Area = $\frac{1}{2} bh$

Circle: Area = πr^2 Circumference = $2\pi r$

Sphere:

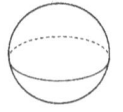

Surface Area = $4\pi r^2$
Volume = $\frac{4}{3}\pi r^3$

Cylinder:

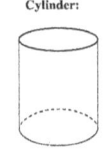

Surface Area = $2\pi rh + 2\pi r^2$
Volume = $\pi r^2 h$

Rectangular Solid:

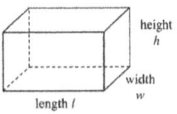

Surface Area = $2lw + 2lh + 2wh$
Volume = lwh

ANSWER SHEET FOR FULL PRACTICE TEST

Question	Answer Choice	Question	Answer Choice	Question	Answer Choice
1.		13.		25.	
2.		14.		26.	
3.		15.		27.	
4.		16.		28.	
5.		17.		29.	
6.		18.		30.	
7.		19.		31.	
8.		20.		32.	
9.		21.		33.	
10.		22.		34.	
11.		23.		35.	
12.		24.		36.	

The following table will divide the questions by chapter. Circle the questions you missed. This will help to identify your strengths and needs and will help focus study time where it is most needed.

Chapter 2	Chapter 3	Chapter 4	Chapter 5	Chapter 6
Understanding Numbers and the Number System	Pre-Algebra	Algebra	Measurement and Geometry	Statistics and Probability
QUESTIONS:	QUESTIONS:	QUESTIONS:	QUESTIONS:	QUESTIONS:
1	3	6	10	11
2	4	8	16	12
22	5	14	18	13
28	7	29	19	15
33	9		23	17
34	20		25	21
	24		27	19
	26		30	26
	32		31	
	35			
	36			

Appendix B

36 multiple choice questions

75 minutes maximum time

Directions: Read each item carefully and choose the best answer response.

1. The price of buying a portable CD player went down from $40 to $10 in the past ten years. What was the percent decrease in the price of a portable CD player in the past ten years?
   (a) 500%
   (b) 75%
   (c) 10%
   (d) $30

2. When the number 4.72×10^{-4} is multiplied by 4, what is the resulting digit in the thousandths place?
   (a) 4
   (b) 7
   (c) 2
   (d) 1

3. Nine children come to a birthday party at a local video arcade. All 9 children want equal time to play *Dino Revenge*, their favorite video game. There are two *Dino Revenge* games at the arcade. The party is from 3:00 to 4:30. How many minutes would each child have to play *Dino Revenge* so that everyone has equal time?
   (a) 10 minutes
   (b) 20 minutes
   (c) 40 minutes
   (d) 50 minutes

4. Which of the following terms is a factor of $a^2b^4c^2+ab^2c^3$?
   (a) ab^2c^2
   (b) a^2bc
   (c) abc^3
   (d) a^4bc

5. Which of the following expressions is equal to $10a(2ab+5)$?
   (a) $20a^2b+50a$
   (b) $70a^2b$
   (c) $20ab+50a$
   (d) $25a^2b$

6. Which of the following is the equation of a line that is perpendicular to the line described by the equation $y=-\frac{1}{2}x + 4$ and has a *y*-intercept of 2?
   (a) $y = -\frac{1}{2}x+2$
   (b) $2y = 4x+4$
   (c) $m = 2$
   (d) $y = -\frac{1}{2}x+4$

7. Which of the following terms is a factor of $a^4b^2c+a^2c^2$?
   (a) a^4bc
   (b) b
   (c) a^2b
   (d) a^2c
8. Which of the following equations listed below represent the given graph?

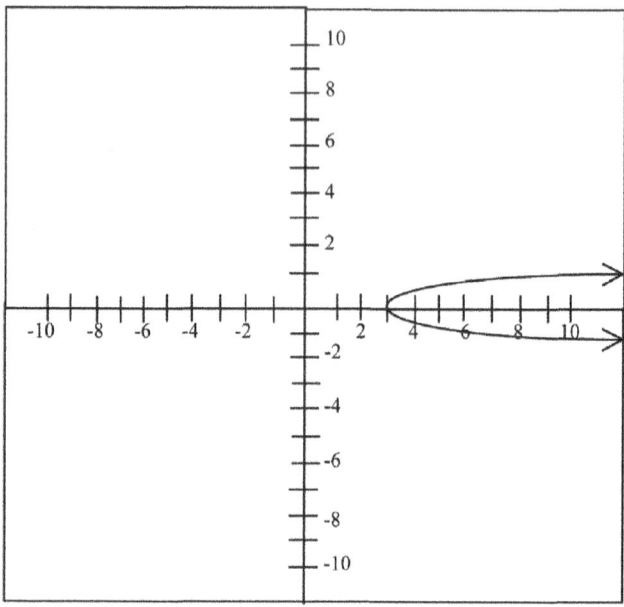

   (a) $x=y^2+3$
   (b) $y=x^2+3$
   (c) $x=-y+3$
   (d) $x=y^2-3$
9. Thirteen less than twice a number is seventeen more than half the number. What is the number?
   (a) $13<2x$
   (b) $13<x<17$
   (c) 20
   (d) 30
10. During Spring Break, the basketball team is traveling 72.5 kilometers in all for a tournament. They already traveled 35 kilometers. How many meters remain before the team reaches their tournament?
    (a) 37,500
    (b) 37.5
    (c) 107.5
    (d) 2,537.5
11. Given the data set 36, 22, 10, 34, and 12, which one of the following numbers can be added to the data so that the median of the six numbers is 26?
    (a) 26
    (b) 30
    (c) 10
    (d) 36

12. A standard die labeled 1–6 is rolled and a coin is flipped simultaneously. What is the total number of possible outcomes that could occur?
    (a) 12
    (b) 2
    (c) 8
    (d) 25
13. A bag contains 8 blue marbles, 4 red marbles, and 2 yellow marbles. A child randomly pulls out one marble from the bag and, without replacing it, takes out a second marble. What is the probability that the child would pull out a blue marble first and a yellow marble second?
    (a) $\frac{8}{14} \times \frac{2}{14}$
    (b) $\frac{1}{8} \times \frac{1}{2}$
    (c) $\frac{8}{14} \times \frac{2}{13}$
    (d) $\frac{8}{14} + \frac{2}{13}$
14. Which of the equations listed below best represents the system of inequality that is given?

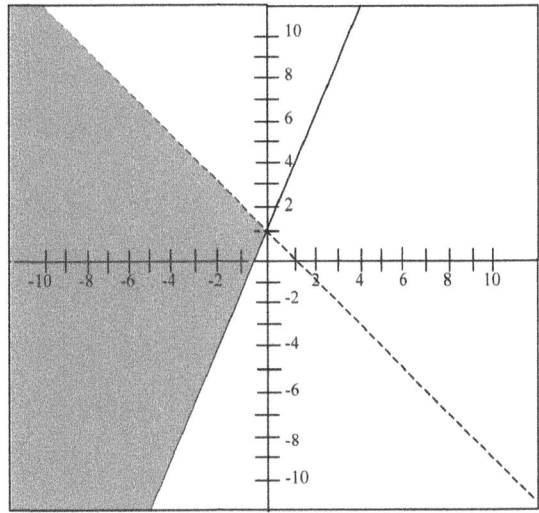

    (a) $y \geq 3x-1$
        $y < x+1$
    (b) $y \leq \frac{1}{3} x +1$
        $y > 2x-1$
    (c) $y \leq 3x+1$
        $y > -x+1$
    (d) $y \geq 3x+1$
        $y < -x+1$
15. Twelve runners are competing for the gold, silver, and bronze Olympic medals. In how many different arrangements can the medals be awarded?
    (a) 36
    (b) 1,320
    (c) 15
    (d) 220

16. Find a counterexample for the following statement:
    All quadrilaterals are parallelograms.
    (a) rectangle
    (b) square
    (c) trapezoid
    (d) rhombus
17. In a survey of members of the Solar Fitness Athletic Club, the following information was collected: 34 members use the basketball courts, 50 members use the weight room, and 29 members use the pool. Ten members use all three services, while 16 of those surveyed use the weight room and the pool, 23 members use the weight room and the basketball courts, and 18 members use the basketball courts and the pool. Fourteen members stated that they do not use any of these services. How many members were surveyed in all? (Hint: Construct a Venn Diagram to help answer the question.)
    (a) 80
    (b) 103
    (c) 68
    (d) 50
18. A rectangular dog park is 20 meters long and 21 meters wide. A walkway for dog owners extends diagonally across the play area. How long is the walkway?
    (a) 29 meters
    (b) 420 meters²
    (c) 82 meters
    (d) 10 meters
19. Which option below could be the length of a rectangle whose perimeter is 60 centimeters if the length is four times the width?
    (a) 10 centimeters
    (b) 80 centimeters
    (c) 120 centimeters
    (d) 24 centimeters
20. A positive number is cubed and then the result is doubled. The final result is 128. What is the number?
    (a) 12
    (b) 64
    (c) 4
    (d) 200
21. Which of the following types of graphs would be best to use to demonstrate a family's budget in a month?
    (a) line graph
    (b) bar graph
    (c) circle graph
    (d) box-and-whisker plot
22. Using the number line below, which of the following expressions results in the greatest value?

(a) $x+y+z$
(b) $-100x$
(c) $10x+3y$
(d) $10(x+z-20)$

23. Two square picnic tables are pushed together to make one larger rectangular picnic table. If each square table has an area of 25 square feet, what is the perimeter, in feet, of the larger rectangular picnic table?
    (a) 50 ft²
    (b) 30 ft
    (c) 40 ft
    (d) 15 ft

24. Twelve years ago, Andy was $\frac{1}{3}$ the age of his brother. Andy is now twenty-four. How old is his brother now?
    (a) 36
    (b) 48
    (c) 72
    (d) 46

25. A rectangular cereal box is $\frac{3}{4}$ full. If the box measures 21 inches, by 18 inches, by 16 inches, how much space is taken up by the cereal?
    (a) 1,512 in³
    (b) 1,514 in²
    (c) 6,048 in³
    (d) 4,534 in³

26. A five-foot ten-inch person standing by a tree casts a ten-foot shadow. The tree casts a sixteen-foot shadow. What is the height in feet and inches of the tree?
    (a) 20 ft. 3 in.
    (b) 9 ft. 4 in.
    (c) 112 in.
    (d) 70 in.

27. If point C below is translated to a new position of C' (5,–5) and every point in triangle ABC follows a similar translation, what are the coordinates of image point A' under this translation?

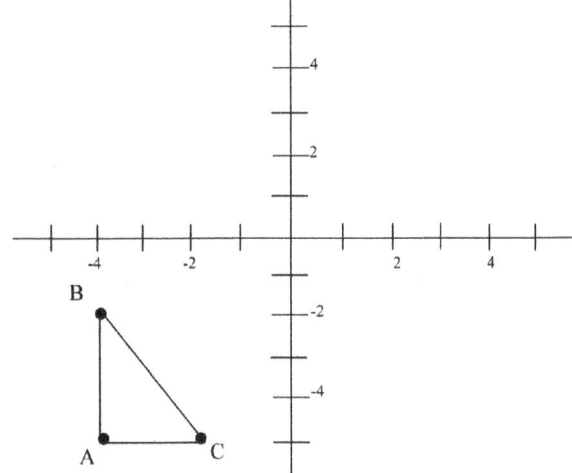

(a) (4,−5)
(b) (3,−5)
(c) (−4,5)
(d) (−4,−5)

28. Which of the following expressions could be utilized to ascertain the value of $5,000 invested at 2.5% interest and compounded annually for ten years?
    (a) $5000(0.025^{10}+1)$
    (b) $5000(10 \times 0.025+1)$
    (c) $5000(1.025)^{10}$
    (d) $5000 \times 10 \times 1.025$

29. In the coordinate plane, what are the coordinates of the midpoint of a line segment connecting point A at (−3,5) and point B at (−3,−1)?
    (a) (10,12)
    (b) (2,−3)
    (c) (−2,3)
    (d) (−3,2)

30. Use the diagram below to answer the question that follows.

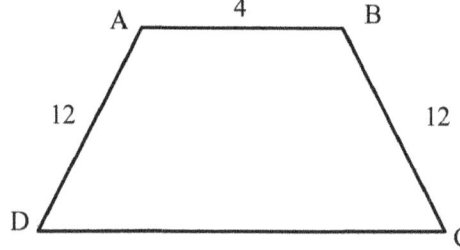
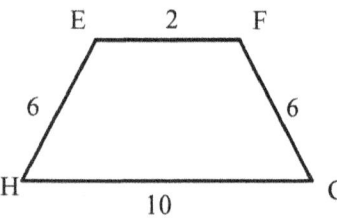

In the drawing above, trapezoid ABCD is similar to trapezoid EFGH. Which of the following proportions will calculate the length of side CD?
(a) $\dfrac{2}{4} = \dfrac{6}{CD}$
(b) $\dfrac{4}{CD} = \dfrac{10}{2}$
(c) $\dfrac{CD}{10} = \dfrac{6}{12}$
(d) $\dfrac{2}{4} = \dfrac{10}{CD}$

31. About how many cubic meters of water will it take to fill a round swimming pool that has a radius of 4 meters and an average depth that is ¾ the length of the radius?
    (a) 48π
    (b) 12π
    (c) 2π
    (d) 24π

32. If a, b, c, and d are variables, which of the following equations is evidence that the following equation forms a proportion?

$$\frac{a}{b} = \frac{c}{d}$$

(a) $a+c=b+d$
(b) $a+d=b+c$
(c) $ad=bc$
(d) $ac=bd$

33. What is the correct order of the following numbers from least to greatest?

$$0.4, \tfrac{1}{2}, 4^{-1}, -0.4, 0.324$$

(a) $0.4, \tfrac{1}{2}, 4^{-1}, -0.4, 0.324$
(b) $-0.4, 4^{-1}, 0.324, 0.4, \tfrac{1}{2}$
(c) $\tfrac{1}{2}, 0.4, 0.324, 4^{-1}, -0.4$
(d) $4^{-1}, -0.4, 0.4, 0.324, \tfrac{1}{2}$

34. A holiday shopper's entire Black Friday purchase included two scarves that cost $10.50 each, a sweater that was on sale for $21.00, and a pair of shoes that cost $49.99. If the sales tax on the entire purchase was 6%, about how much change did the shopper receive from a $100 bill?
    (a) $1.40
    (b) $5.52
    (c) $2.70
    (d) $2.49

35. At the neighborhood 4-H sale, a local farmer bought 40 vegetable plants. Some cost $2 per plant and the rest cost $4 per plant. The farmer paid $200 in all. If p represents the number of $4 plants, which of the following equations could be solved to find the number of $4 plants purchased?
    (a) $4p+2(40-p)=200$
    (b) $p(2+4)=200$
    (c) $2p+4(40-p)=200$
    (d) $4(p-40)=200$

36. The visual below describes the geometric relationship for the following procedure in algebra?

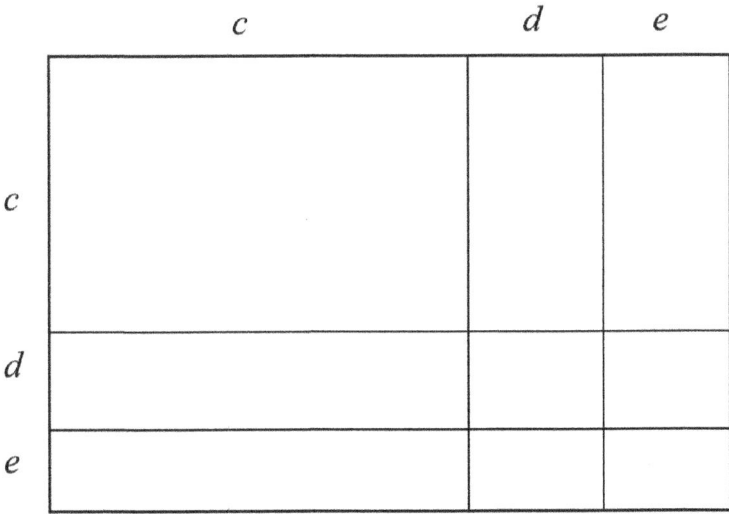

(a) simplifying an expression
(b) factoring a binomial
(c) using the distributive property
(d) squaring a trinomial

ANSWERS AND DETAILED EXPLANATIONS FOR FULL PRACTICE TEST 2

1. Answer: (b) 75%
   Estimate your answer. $10 is one-fourth of $40. The percent of decrease should be 100%−25%, which is a decrease of 75%.
   Subtract the amount of decrease: $40−$10=$30. Divide the amount of decrease by the original amount of the CD player now: 30÷40=0.75. Convert 0.75 to a percent by multiplying it by 100. The percent decrease of buying a portable CD player is 75%. Check the estimate to see if the answer makes sense. Yes, the answer is reasonable.
   Answer choice (a) is too high of a percent decrease. Answer choice (c) doesn't make sense because it is too low of a percent decrease. Answer choice (d) does not answer the question. $30 is the amount of decrease not the percent of decrease.
   Topic of Question 1: Percentages. More on percentages can be found in chapter 2.

2. Answer: (d) 1
   Multiply 4.72×10^{-4}=0.000472 (or 4.72E−4). Whenever multiplying by a negative power of ten, move the decimal point to the left as many places as the exponent indicates. In this case, move the decimal point four places to the left and insert zeroes as place holders. Multiply 0.000472×4=0.001888. The digit in the thousandths place is 1. None of the other answer choices make sense because the product of $4.72 \times 10^{-4} \times 4$ does not include a 4, 7, or 2 in the answer.
   Topic of Question 2: Different representations of numbers. More on writing numbers in different forms can be found in chapter 2.

3. Answer: (b) twenty minutes
   Convert the time of the party (3:00–4:30) to total minutes: 1.5 hours = 90 minutes. Multiply the total number of minutes by the number of available *Dino Revenge* games: 90 minutes × 2 *Dino Revenge* games=180 total minutes of game use. Divide the total of video minutes by the number of children who want to play: 180 minutes÷9 children=20 minutes. Each child can spend 20 minutes playing *Dino Revenge*. Choice (a) will leave unused minutes. Choices (c) and (d) do not allow every student to have equal access to the game, because 180 minutes cannot be evenly divided by 40 or 50.
   Topic of Question 3: Translating word problems as expressions. More on algebraic principles can be found in chapter 3.

4. Answer: (a) ab^2c^2
   a is a factor of both a^2 and a. b^2 is a factor of both b^4 and b^2. c^2 is a factor of both c^2 and c^3. Answer choice (b) doesn't work because a^2 is not a factor of the second term. Answer choice (c) is incorrect because c^3 is not a factor of the first term. Answer choice (d) is incorrect because a^4 is not a factor of either term.
   Topic of Question 4: Factoring. More on factoring can be found in chapter 3.

5. Answer: (a) $20a^2b+50a$
   The terms in the parentheses are unlike terms so they cannot be combined. Use the distributive property to multiply ($10a \times 2ab$ to get $20a^2b$) and ($10a \times 5$ to get

Appendix B

$50a$). The result is $20a^2b+50a$. Answer choice (b) is incorrect because the terms cannot be combined in the parentheses. Answer choice (c) is wrong because $10a \times 2ab = 20a^{1+1=2}b$. Answer choice (d) is wrong because $20a^2b$ cannot be combined with 5 to make $25a^2b$ because they are unlike terms.

Topic of Question 5: Applying the distributive property. More on the distributive property can be found in chapter 3.

6. Answer: (b) $2y=4x+4$

    Linear equations in slope intercept form are perpendicular if their slopes are negative reciprocals. $y = 2x + 2$ is a line perpendicular to $y = -\frac{1}{2}x + 4$ because the slopes are negative reciprocals to each other.

    A line in the form $y=mx+b$ is in slope-intercept form where m is the slope and b is the y-intercept. Convert each of the given equations to slope-intercept form.

    Answer choice (c) is the slope. It is not an equation of a line and therefore does not answer the question. It can be automatically eliminated.

    Answer choice (a) $y=-\frac{1}{2}x+2$ is in slope-intercept form already. It has the same slope as the given equation. Therefore, they form parallel lines, not perpendicular lines.

    In answer choice (b), first solve $2y=4x+4$ for y by dividing each term in the equation by 2 to get $y=2x+2$.

    The y-intercept of $y=2x+2$ is 2. This equation is perpendicular to the given equation and has a y-intercept of 2.

    Answer choice (d) has the correct slope. However, it has a y-intercept of 4.

    Topic of Question 6: Equation of perpendicular lines. More on relating linear equations can be found in chapter 4.

7. Answer: (d) a^2c

    a^2 is a factor of both a^4 and a^2. c is a factor of both c^2 and a^2. So a^2c is a factor of $a^4b^2c+a^2c^2$. Answer choice (a) is wrong because both a^4 and b are not factors of both terms. Answer choice (b) is not a factor of the second term. Answer choice (c) is not a factor of the second term because b is not a factor of a^2c^2.

    Topic of Question 7: Factoring. More on factoring can be found in chapter 3.

8. Answer: (a) $x = y^2+3$

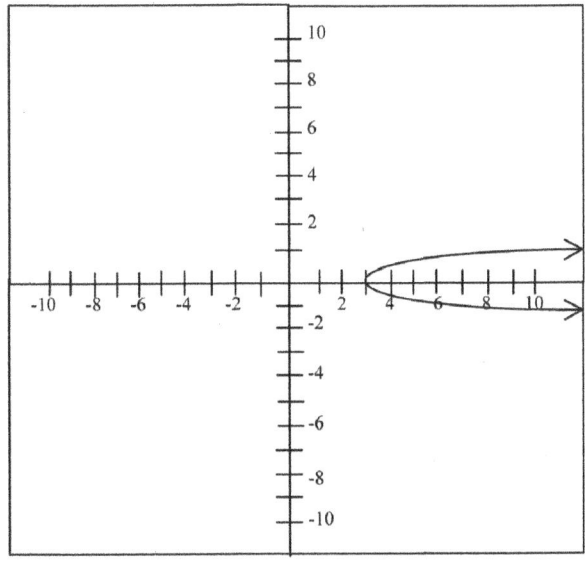

First, it is easy to see that answer choice (b) $y=x^2+3$ can be eliminated because the given graph shows a horizontal parabola, with an equation of $x=y^2$ instead of a vertical parabola with an equation of $y=x^2$. Answer choice (c) $x=-y+3$ can also be eliminated because y is not squared.

The equation for a sideways parabola is $x=y^2$. A slide to the right means that the equation will have a positive 3. The result is (a) $x=y^2+3$. Check the answer by replacing the vertex for x and y: $3=0^2+3$, $3=3$. Answer choice (a) is correct.

Topic of Question 8: Graphing parabolas. More on this can be found in chapter 4.

9. Answer: (c) 20

Interpret the language of the equation from words to math language with numbers, operations, and symbols. The easiest way to do this is to picture what the language would look like as an equation. It is helpful to keep reading the description until the equation reads exactly like the written interpretation.

Thirteen less than twice a number can be rewritten as $2x-13$.

Continue on with the rest of the description: "Thirteen less than twice a number *is* seventeen more than half the number" can be interpreted in the language of math as: $2x-13=\frac{1}{2}x+17$ The word *is* describes an equation and not just an expression.

To make sure the interpretation is correct, read over both written and math descriptions to make sure they say the same thing. Then, solve the equation for x.

$$2x-13=\tfrac{1}{2}x+17$$

Add 13 to both sides of the equation.

$$2x=30+\tfrac{1}{2}x$$

Subtract $\tfrac{1}{2}x$ from both sides of the equation.

$$1\tfrac{1}{2}x=30$$

Divide each side of the equation by $1\tfrac{1}{2}$.

$$x=20$$

Replace 20 back into the original equation:

$$2(20)-13=\tfrac{1}{2}(20)+17$$
$$27=27$$

The answer is correct.

Answer choices (a) and (b) convert the language of *less than* incorrectly as *is less than*. The concept of *less than* does not represent an inequality. Both (a) and (b) can be eliminated. Answer (d) is an incorrect calculation.

Topic of Question 9: Translating words into algebraic expressions. More on translating words into algebraic expressions can be found in chapter 3.

10. Answer: (a) 37,500

The total distance for the team to travel is 72.5 kilometers. They already traveled 35 kilometers. To determine how much of the trip is left, find the difference: 72.5−35=37.5 kilometers. Now convert the kilometers to meters by multiplying 37.5 × 1,000 because there are 1,000 meters in one kilometer. 37.5 × 1,000=37,500 meters are still left for the basketball team to travel. Answer choice (b) is wrong because the answer is still in kilometers. Answer choice (c) is incorrect because it is the sum of the two measures, which doesn't answer the question. Answer choice (d) is the product of the two measures. This answer choice doesn't make sense at all and can be eliminated immediately.

Topic of Question 10: Estimating and measuring in the metric system. More on the metric system can be found in chapter 5.

11. Answer: (b) 30

The median is the middle number of a data set when the numbers are in order from least to greatest. First, order the numbers from least to greatest.

The median with the new number added would have to be between 22 and 34 in order to change the median to 26. Choices (a) and (b) are the only answers that could work:

10, 12, **22**, **26**, 34, 36—the median is the middle number between 22 and 26, which is 24. This is not the correct median.

10, 12, **22**, **30**, 34, 36—the median is the middle number between 22 and 30, which is 26. This is the correct median.

Topic of Question 11: Measures of central tendency. More on mean, median, and mode can be found in chapter 6.

12. Answer: (a) 12

The Fundamental Counting Principle states that if there are *m* ways for one event to occur, and *n* ways for a second event to occur, then there are *m* × *n* ways for both to occur. There are six faces of the cube, each with a different number, and only two possible outcomes for the coin: heads or tails. Then, according to the Fundamental Counting Principle, there are 6 × 2=12 possible combinations that can occur when flipping a coin and rolling a die. Answer choice (b) in incorrect because it only describes the number of items used in the experiment—a die and a coin. Answer choices (c) and (d) do not make any sense in the problem and can be eliminated.

Topic of Question 12: The Fundamental Counting Principle. More on this can be found in chapter 6.

13. Answer: (c) $\frac{8}{14} \times \frac{2}{13}$

There are 8 blue marbles out of the 14 marbles in all. There is an 8 in 14 probability the child randomly picks a blue marble on the first try. Because the first marble is not replaced, there are now only 13 marbles in the bag, so there is a 2 in 13 probability to get a yellow marble on the second try. The probability of one outcome *and* another outcome occurring requires multiplication. Answer choice (a) is incorrect because it would be the answer if the first marble were put back in the bag before the second one was picked, that is, "with replacement." Answer choice (b) is incorrect because it is not a reasonable answer to the problem and can be eliminated. Answer choice (d) is wrong because it would be the probability of getting a blue marble *or* getting a yellow marble without replacement.

Topic of Question 13: Probability. More on probability can be found in chapter 6.

14. Answer: (d) $y \geq 3x+1$

$y < -x+1$

The first inequality graphed has a positive slope and a y-intercept of 1, and all points on the line are part of the solution set. The slope of the first line is 3. The second line has a negative slope, all points on the line are not part of the solution set, and it also has a y-intercept of 1. The slope of the second line is -1. Therefore, the first inequality represents a line slanting upward that is solid and the second inequality represents a line slanting downward that is dashed. Answer choice (d) is the only choice that meets all the above criteria. Now choose a point in the shaded region to check to see if it is correct. Choose an easy point like $(-10,0)$.

$$y \geq 3x+1$$
$$0 \geq 3(-10)+1$$
$$0 \geq -29$$
$$y < -x+1$$
$$0 < -(-10)+1$$
$$0 < 11$$

Answer choice (d) is correct

Topic of Question 14: Graphing inequalities. More on this can be found in chapter 4.

15. Answer: (b) 1,320

A permutation is an arrangement or listing is which *order is important*. To find the permutation of n objects taken r at a time, use the following formula:

$$_nP_r = \frac{n!}{(n-r)!}$$

$$_{12}P_3 \frac{12!}{(12-3)!} = \frac{12 \cdot 11 \cdot 10 \cdot 9!}{9!}$$

Simplify the fraction to get $12 \times 11 \times 10 = 1{,}320$ different ways to arrange who might get the gold, silver, or bronze medal for the twelve runners. Answer choice (a) is incorrect. It is the product of the twelve runners and the three medals. There are many more arrangements than 36. Answer choice (c) is not a reasonable answer either and can be eliminated as an option. Answer choice (d) is the number of combinations possible. Combinations do not consider order, and when awarding medals, order is very important.

Topic of Question 15: Permutation. More on permutations and combinations can be found in chapter 6.

16. Answer: (c) Trapezoid

A quadrilateral is a polygon with four sides. A parallelogram is a quadrilateral whose opposite sides are parallel and of the same length.

A counterexample will make a statement untrue. Try each answer choice to see which example makes the statement false. Answer choices (a), (b), and (d) are all true: rectangles, squares, and rhombuses are all parallelograms by definition. A trapezoid is a quadrilateral that has exactly one set of parallel sides. It is not a parallelogram by definition.

Topic of Question 16: Analyzing polygons using attributes. More on polygons can be found in chapter 5.

17. Answer: (a) 80

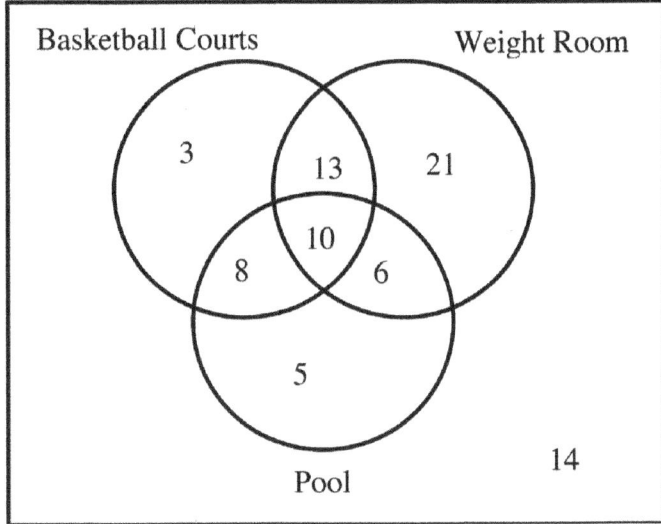

Answer choice (d) 50 can be automatically eliminated because there are 50 members who use the weight room, without any regard to those who use the basketball courts and the pool.
Construct a Venn Diagram describing the results of the survey. Always begin with the innermost section of the Venn Diagram—in this case the 10 members who use all three services. Then work out and complete the sections where members use two of the three services by subtracting 10 from those numbers (23−10=13 use both the basketball courts and the weight room, 16−10=6 use the weight room and the pool, and 18−10=8 use both the basketball courts and the pool). In order to calculate the number of members who only use one service at Solar Fitness, add each section within each circle that represents one service filled in so far and subtract that from the number of members using each service (34−31=3 members use only the basketball courts, 50−29=21 members use only the weight room, and 29−24=5 members use only the pool). Add all of the sections, as well as the 14 members who do not use any of the services at Solar Fitness, to determine the total number of members surveyed (3+13+21+8+10+6+5+14=80 members were surveyed).
Topic of Question 17: Venn Diagrams. More on Venn Diagrams can be found in chapter 6.

18. Answer: (a) 29 meters
The dog park is a rectangle split in the middle with a diagonal walkway. This divides the rectangle into two right triangles. To calculate the length of the diagonal, use the Pythagorean Theorem to solve the problem: In any right triangle, the sum of the squares of the lengths of the legs is equal to the square of the length of the hypotenuse. Estimate your answer. Answer choice (d) does not make sense because the opposite side of the right angle (or hypotenuse) of a right triangle is the longest side of the triangle. Answer choice (b) is the area of the rectangle and therefore is incorrect because it doesn't answer the question. Answer choice (c) is the perimeter of the rectangle. This too does not answer the question and can

be eliminated as a possible answer. The only answer that really makes sense is answer choice (a). Do the calculations to check your estimate.

$$c^2=a^2+b^2$$

The diagonal walkway represents the hypotenuse of the triangle.

$$c^2=20^2+21^2$$

$$c^2=400+441$$

Calculate the square root of 841.
The hypotenuse = 29 meters.
The diagonal walkway is 29 meters long.
This answer makes sense with the original estimate.
Topic of Question 18: The Pythagorean Theorem. More on estimating and calculating measurements using the Pythagorean Theorem can be found in chapter 5.

19. Answer: (d) 24 centimeters
The perimeter of the rectangle is 60 centimeters. Answer choices (b) and (c) are automatically eliminated because the length of one side cannot be longer than the distance around the entire rectangle. Answer choice (a) doesn't make sense either because the length is four times the width, so a length of 10 centimeters would make the width 2.5 centimeters. These dimensions would not result in a perimeter of 60 centimeters. The only answer that makes sense is answer choice (d). Now do the calculations to check your answer.
Set up an equation for perimeter.

$$2w+2(4w)=60$$

Multiply and collect like terms.

$$10w=60$$

Divide by 10.

$$w=6$$

The width is 6 centimeters. Since the length is four times the width, the length is 24 centimeters.
Replace w in the equation to check the answer. $2(6)+2(4\times6)=60$. The original estimate is correct.
Topic of Question 19: Applying formulas to calculate and determine lengths. More on using formulas to calculate lengths can be found in chapter 5.

20. Answer: (c) 4
Create an algebraic equation from the problem given: $2(x^3)=128$. Using this equation automatically eliminates answer choice (d) because the product of a positive number cannot be smaller than one of its factors. Answer choice (b) doesn't make

Appendix B

sense either because cubing 64 would be much larger than 128, even without doubling it. Now use the equation to solve the problem.

$$2(x^3)=128$$

Divide both sides of the equation by 2.

$$x^3=64$$

Find the cube root of 64.

$$x=4$$

Substitute 4 into the original equation: $2(4^3)=128$. Answer choice (c) is correct.
Topic of Question 20: Using variables in equations. More on creating and solving equations from word problems can be found in chapter 3.

21. Answer: (c) circle graph
Answer choice (c) is correct because a circle graph is best used to compare part to the whole and parts to other parts. A circle graph would best display how the family uses their funds during a month. Answer choice (a) is incorrect because a line graph is a graph that shows changes in data over time. Answer choice (b) is incorrect because a bar graph is a graph used to show relationships or comparisons between categories. Answer choice (d) is incorrect because a box-and-whisker plot shows the distribution and spread of data.
Topic of Question 21: Choosing appropriate ways to show data. More on appropriateness of graphs can be found in chapter 6.

22. Answer (b) $-100x$
First identify the interval (space between the tick marks). The interval is 1.
Find the value of each of the variables: $x=-3, y=1, z=3$.
Substitute the values of each of the variables and evaluate each expression.
(a) $x+y+z=-3+1+3=1$
(b) $-100x =-100 \times -3=300$
(c) $10x+3y=10(-3)+3(1)=-30+3=-27$
(d) $10(x+z+-20)=10(-3+3+-20)=10(-20)=-200$
Answer choice (b) 300 is the greatest value.
Topic of Question 22: Relative magnitude of real numbers. More on real numbers can be found in chapter 2.

23. Answer: (b) 30 feet
The area of a smaller picnic table is 25 ft². The formula provided for calculating area is area $= lw$. Since it is a square, the sides are the same. Calculate the square root of 25. Each side is 5 feet.
The two tables are pushed together giving the rectangular table a length of 10 feet and a width of 5 feet. The formula for perimeter of a rectangle is $P=2l+2w$. Substitute the values for length and width: $P=2(10)+2(5)$. The perimeter is 30 feet.
Answer choice (a) can be eliminated since it does not answer the question. It is the area of the rectangle. Perimeter is not measured in square units. Answer choice

(c) is incorrect because 10 feet is the perimeter of one of the square picnic tables. Answer choice (d) is only half the perimeter.
Topic of Question 23: Applying formulas to calculate and determine lengths. More on using formulas to calculate lengths can be found in chapter 5.

24. Answer: (b) 48

Andy is now 24. Twelve years ago, Andy was $\frac{1}{3}$ the age of his brother. Subtract 12 from Andy's present age: 24−12=12. Multiply Andy's age twelve years ago by 3 because at that time Andy was $\frac{1}{3}$ the age of his brother. 3 × 12=36. Twelve years ago Andy's brother was thirty-six years old. Add 12 to his age to find the age of Andy's brother now: 36+12=48. Andy's brother's present age is 48.

Answer choice (a) is wrong because it is the brother's age 12 years ago. Answer choice (c) is wrong because it does not answer the question. 72 is three times Andy's present age.

Topic of Question 24: Applying algebraic principles in word problems. More on algebraic principles can be found in chapter 3.

25. Answer: (a) 1,512 in³

Space indicates volume. Calculate the volume of the rectangular solid. Volume=*lwh*. Substitute the values in the formula. V=21 × 18 × 16. The volume of the cereal box is 6,048 in³. The cereal fills $\frac{1}{4}$ the volume of the box. Divide 6,048 by 4.

$$6048 \div 4 = 1512$$

The cereal takes up 1,512 in³ of the box.

Answer choice (b) can be immediately eliminated because volume is measured in cubic units. Answer choice (c) gives the total volume of the box. It does not answer the question. Answer choice (d) is wrong because 4,534 in³ is the volume of the box that is *not* filled with cereal.

Topic of Question 25: Applying formulas to calculate and determine volume. More on using formulas to calculate volume can be found in chapter 5.

26. Answer: (b) 9 ft 4 in

(a) First set up a proportion to represent the given information:

$$\frac{5 \text{ft } 10 \text{in}}{10 \text{ ft}} = \frac{x}{16 \text{ ft}}$$

Convert the height of the person to inches to make the calculations easier: 5 × 12=60+10=70 inches.

The person is 70 inches tall. Cross multiply to solve for *x*.

$$10x = 70 \times 16$$
$$10x = 1120$$

Divide by 10. The tree is 112 inches in height.
Convert the inches back into feet by dividing 112 by 12.
112÷12=9$\frac{1}{3}$ feet, which is 9 ft 4 in.

Answer choice (a) can be immediately eliminated because the tree cannot be longer than its shadow if the person standing near the tree is shorter than its shadow. Answer choice (c) is wrong because it is the answer in inches. The question asks

for the answer converted to feet and inches. Answer choice (d) is the height of the person in inches. It does not answer the question and can be eliminated as a possible choice.
Topic of Question 26: Applying algebraic principles in proportions. More on algebraic principles in proportions can be found in chapter 3.

27. Answer: (b) (3,−5)

A translation is a slide. Point C is sliding right from coordinates (−2,−5) to the image point C' (5,−5). That is a slide seven units to the right. Consequently, all the translation image *x*-coordinates will increase by seven and the *y*-coordinates will stay the same. Point A under this translation will be A' (3,−5). Answer choice (a) is incorrect because (4,−5) is the image of point being reflected over the *y*-axis. Answer choice (c) is incorrect because (−4,5) is the image of point A being reflected over the *x*-axis. Answer choice (d) is incorrect because it is the coordinates of point A before the translation.
Topic of Question 27: Displaying transformations. More on transformations can be found in chapter 5.

28. Answer: (c) $5000(1.025)^{10}$

The principal invested is $5,000. The rate of the investment is 2.5%, which is equivalent to 0.025 as a decimal. The interest is compounded annually over 10 years. Use the formula provided for compound interest: $A=P(1+r)^t$ where P=principal ($5,000), r=rate (0.025) and t=time (10). Answer choice (c) $5000(1.025)^{10}$ is the only choice that represents each value appropriately.
Topic of Question 28: Compound interest. More on compound interest can be found in chapter 2.

29. Answer: (d) −3,2

To calculate the midpoint of a line segment given two endpoints, use the formula provided:

$$\left(\frac{x_1+x_2}{2}, \frac{y_1+y_2}{2}\right)$$

Substitute each of the point values and calculate the midpoint. The order of the endpoints does not matter.

$$\frac{-3+-3}{2} = \frac{-6}{2} = -3$$

The *x*-coordinate is −3.

$$\frac{5+-1}{2} = \frac{4}{2} = 2$$

The *y*-coordinate is 2.
The midpoint is (−3,2).
Answer choices (a) through (c) are incorrect because of wrong calculations.
Topic of Question 29: Identifying the midpoint of a line segment. More on this can be found in chapter 4.

30. Answer: (d) $\frac{2}{4} = \frac{10}{CD}$

Since trapezoid ABCD and trapezoid EFGH are similar, their corresponding sides form proportions. It is important to first match up the corresponding sides for the trapezoids. Side AB corresponds to side EF. Side BC corresponds to side FG.

Side CD corresponds to side GH. Side DA corresponds to side HE. Therefore the following proportions are correct:

$$\frac{2}{4} = \frac{6}{12} = \frac{6}{12} = \frac{10}{CD}$$

Answer choice (c) is the only option where the ratios are equal. It is evident as well that trapezoid EFGH is one-half the size of trapezoid ABCD.

Topic of Question 30: Applying the relationship of corresponding sides of similar figures. More on this can be found in chapter 5.

31. Answer: (a) 48π

    A round swimming pool is a cylinder. Calculate the volume of the cylinder to find how much water it will take to fill the pool. Use the following formula provided for volume of a cylinder:

    $$V=\pi r^2 h$$

    Substitute the known values in the equation. $r = 4$ meters, and height (described in the problem as depth) = $\frac{3}{4}$ of the radius, which is equal to $\frac{3}{4} \times 4 = 3$ meters.

    $$V = \pi \times 4^2 \times 3$$

    Simplify the exponent.

    $$V = \pi \times 16 \times 3$$
    $$V = 48\pi$$

    Answer choice (b) is wrong because the radius is not squared. Answer choice (c) is not a reasonable answer because the volume cannot be less than the radius. This answer can be automatically eliminated. Answer choice (d) shows incorrect calculations when simplifying the exponent.

    $$4^2 \neq 8$$
    $$4^2 = 4 \times 4 = 16$$

    Topic of Question 31: Applying formulas to calculate and determine volume. More on using formulas to calculate volume can be found in chapter 5.

32. Answer: (c) $ad=bc$

    Whenever a set of fractions form a proportional relationship, their cross-products are equal. Therefore, a test for proportionality is to multiply and then compare the cross products. Answer choice (c) $ad=bc$ is an accurate way to prove proportionality.

    Answer choices (a) and (b) are not accurate methods for testing proportionality. Answer choice (d) is incorrect because it does not use cross-products as the factors.

    Topic of Question 32: Applying algebraic principles in proportions. More on algebraic principles in proportions can be found in chapter 3.

33. Answer: (b) $-0.4, 4^{-1}, 0.324, 0.4, \frac{1}{2}$
Convert each of the numbers to a decimal. It is usually easier to compare decimals.

$$4^{-1} = \frac{1}{4} = 0.25$$
$$\frac{1}{2} = 0.5$$

Now compare decimals. The negative decimal is the least value. Then compare the positive decimals by comparing each place value one at a time. That means 0.25<0.324 because two-tenths is less than three-tenths. Continue comparing decimal numbers. The numbers from least to greatest are as follows:

$$-0.4, 0.25, 0.324, 0.4, 0.5$$

Convert the numbers back to the original form and rewrite the order from least to greatest: $-0.4, 4^{-1}, 0.324, 0.4, \frac{1}{2}$

Topic of Question 33: Demonstrate understanding of real numbers. More on real numbers can be found in chapter 2.

34. Answer: (d) $2.49
(a) First, calculate the total cost of the purchases.

$$(2 \times \$10.50) + \$21.00 + \$49.99 = \$91.99$$

Find 6% of $91.99 by multiplying .06 × 91.99=5.5194. Round to the nearest cent: $5.5194 rounded to the nearest cent is $5.52.
Add the tax to the total cost of the purchase: $91.99+$5.52=$97.51.
Subtract $97.51 from the $100 bill to find how much change the shopper received.

$$\$100.00 - \$97.51 = \$2.49 \text{ change}$$

Answer choices (a) and (c) do not make sense once the total is calculated with the tax. The change would have to be $2 and some change. Answer choice (b) is the cost of the tax. It can be eliminated because it does not answer how much change the shopper receives.

Topic of Question 34: Display fluency in computation. More on computation of real numbers can be found in chapter 2.

35. Answer: (a) $4p+2(40-p)=200$
If the farmer bought p plants that cost $4 each and 40 plants in all, then the farmer bought $(40-p)$ plants that cost $2 each. The cost of the $4 plants in all is $4p$. The cost of the $2 plants in all is $2(40-p)$. The total cost is $200. Therefore, the equation $4p+2(40-p)=200$ is an appropriate equation to discover the number of $4 plants purchased.
Answer choices (b) and (d) are not reasonable answers and can be eliminated. Answer choice (c) is an equation that would solve the cost of $2 plants if the farmer bought p plants that cost $2.

Topic of Question 35: Applying algebraic principles in word problems. More on algebraic principles in word problems can be found in chapter 3.

36. Answer: (d) squaring a trinomial
    Identify that the figure represents a length and width that includes the same three terms (*c, d,* and *e*). Therefore, geometrically they are each being multiplied by itself. Answer choice (d) squaring a trinomial is the correct choice.
    Topic of Question 36: Distributive property. More on displaying distribution can be found in chapter 3.

Appendix C

Full Practice Test 3

Formulas you will need for the Math Test

Permutation: $_nP_r = \frac{n!}{(n-r)!}$

Combination: $_nC_r = \frac{n!}{(n-r)!r!}$

Simple Interest: $I = P \times r \times t$

Compound Interest: $A = P(1 + r)^t$

Finding the midpoint: $\left(\frac{x_1+x_2}{2}, \frac{y_1+y_2}{2}\right)$

Calculating the distance between two points: $\sqrt{(x_2 - x_1)^2 + (y_2 - y_1)^2}$

Pythagorean Theorem: $c^2 = a^2 + b^2$

Rectangle: Area= lw Perimeter: $2l + 2w$

Triangle: Area = ½ bh

Circle: Area = πr^2 Circumference = $2\pi r$

Sphere:

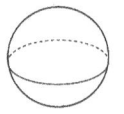

Surface Area = $4\pi r^2$
Volume = $\frac{4}{3}\pi r^3$

Cylinder:

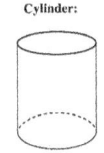

Surface Area = $2\pi rh + 2\pi r^2$
Volume = $\pi r^2 h$

Rectangular Solid:

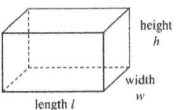

height h
width w
length l

Surface Area = $2lw + 2lh + 2wh$
Volume = lwh

ANSWER SHEET FOR FULL PRACTICE TEST

Question	Answer Choice	Question	Answer Choice	Question	Answer Choice
1.		13.		25.	
2.		14.		26.	
3.		15.		27.	
4.		16.		28.	
5.		17.		29.	
6.		18.		30.	
7.		19.		31.	
8.		20.		32.	
9.		21.		33.	
10.		22.		34.	
11.		23.		35.	
12.		24.		36.	

The following table will divide the questions by chapter. Circle the questions you missed. This will help to identify your strengths and needs and will help focus study time where it is most needed.

Chapter 2	Chapter 3	Chapter 4	Chapter 5	Chapter 6
Understanding Numbers and the Number System	Pre-Algebra	Algebra	Measurement and Geometry	Statistics and Probability
QUESTIONS:	QUESTIONS:	QUESTIONS:	QUESTIONS:	QUESTIONS:
1	2	10	4	9
3	5	21	8	11
14	6	22	17	12
18	7		20	13
25	19		23	15
26	27		24	16
31	30		28	
32	33		29	
35	34			
	36			

Appendix C

36 multiple choice questions

75 minutes maximum time

Directions: Read each item carefully and choose the best answer response.

1. The finale of a TV singing competition had a viewing audience of 14.342 million people, versus its usual audience of 10.1 million people. What was the percent increase for the finale?
   (a) 4.242%
   (b) 4.242
   (c) 100%
   (d) 42%
2. A baseball coach wants to take his little league team to the batting cages to practice hitting. There are 12 players on the team and three batting cages available to reserve. The coach has reserved them for two hours. If the coach wants to make sure each player gets equal time, how many minutes would each player get to practice in the batting cages?
   (a) 30 minutes
   (b) 10 minutes
   (c) 12 minutes
   (d) 50 minutes
3. When (6.25×10^3) is multiplied by (3.1×10^4), what is the resulting digit in the millions place?
   (a) 6
   (b) 7
   (c) 3
   (d) 0
4. Use the diagram below to answer the question that follows:

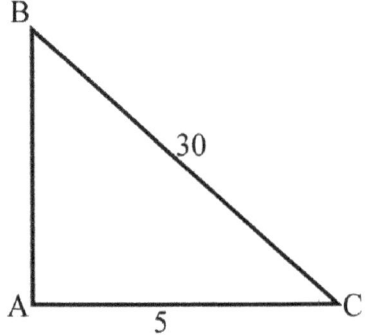 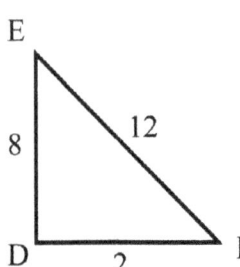

In the drawing above, triangle ABC is similar to triangle DEF. What is the length of side AB?

(a) 20"
(b) 32"
(c) 10"
(d) 8"

5. Which of the following terms is a factor of $x^2y^3z+xy^2z^2$
   (a) xyz^2
   (b) x^2y^2z
   (c) xy^3z
   (d) xy^2z

6. Which of the following expressions is equal to $12b(2bc+6)$?
   (a) $96b^2c$
   (b) $30b^2c$
   (c) $24b^2c+6$
   (d) $24b^2c+72b$

7. The second angle of a triangle is twice as big as the first. The third angle is 50° less than the second. Find the measures of the three angles.
   (a) 180°
   (b) 50°, 100°, 100°
   (c) 50°
   (d) 46°, 92°, 42°

8. A family is traveling on vacation. They have traveled 45 kilometers so far. They have 27,500 meters left to travel. How many kilometers is the trip in all?
   (a) 27,545
   (b) 72.5
   (c) 72,500
   (d) ≈ 611

9. An average grade of 90 is necessary to achieve an "A" for Math 114. A student has scored a 94, 88, 80, and 90 on the first four assessments. What is the lowest score the student can earn on the fifth test to get an "A" for the class?
   (a) 98
   (b) 88
   (c) 84
   (d) 92

10. Which of the following equations listed below represents the given graph?
    (a) $y=x^2+5$
    (b) $y=-x^2+5$
    (c) $y=-x^2-5$
    (d) $y=-x+5$

11. The campus cafeteria randomly selects roast beef, chicken, or fish as the main entrée for students on Sundays. They offer fives sides as well: a small salad, corn, potatoes, carrots, or broccoli. Students are randomly given one side to go with their main choice. The meal also comes with a dessert item. The dessert options on Sunday are frozen yogurt, mixed fruit, or pie. How many different meal combinations are available for students for Sunday dinner?

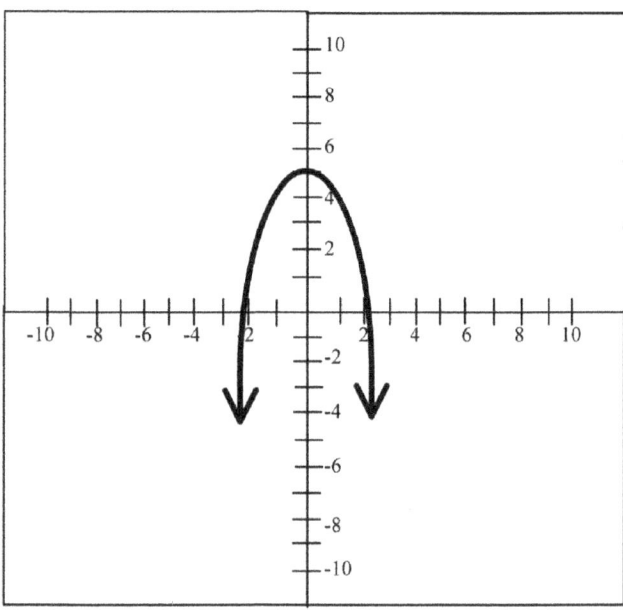

(a) 3
(b) 11
(c) 45
(d) 100

12. A number cube whose faces are marked 3, 6, 9, 12, 15, and 18 is rolled two times. What is the probability of rolling an even number the first roll *and* a number between 1 and 10 the second roll? Identify how to solve this problem.
    (a) $\frac{1}{2} \times \frac{1}{2}$
    (b) $\frac{1}{6} \times \frac{1}{6}$
    (c) $\frac{1}{2} + \frac{1}{2}$
    (d) 2×2

13. There are 8 entries in the Kennel Club National Dog Competition. A first-, second-, and third-place ribbon will be awarded for best of show. In how many ways can the ribbons be awarded?
    (a) 8
    (b) 336
    (c) 24
    (d) 56

14. Find a counterexample for the following statement: *If a number is multiplied by −1, then the result is less than the number.*
    (a) (−2)(−1)
    (b) (5)(−1)
    (c) ($\frac{1}{4}$)(−1)
    (d) (100)(−1)

15. In the Venn Diagram below, the circles represent the viewing behavior of sports fans. The circle labeled F represents those fans who watch football, the circle labeled B represents the fans who watch baseball, and the circle labeled H represents the fans who watch hockey.

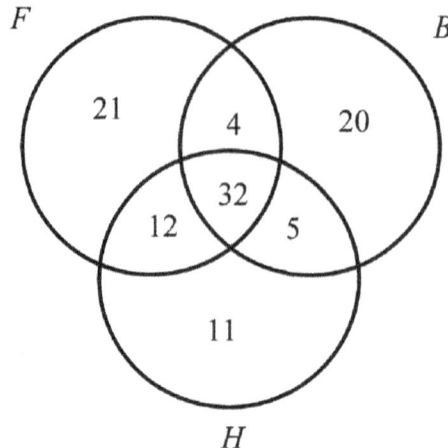

Which statement does not represent the viewing behavior of sports fans based on the above Venn Diagram?
(a) Most sports fans watch more than one sport.
(b) Football is the most viewed sport.
(c) Most fans enjoy watching football and hockey, but not baseball.
(d) Hockey is the least-viewed sport.

16. Which of the following types of graphs would be best to use to highlight and describe the reading grades of 30 elementary students?
(a) line graph
(b) bar graph
(c) circle graph
(d) box-and-whisker plot

17. The perimeter of a rectangle is 40 units with a length of $(2x-1)$ and the width is $(x+6)$. What is the area of the rectangle?
(a) 5 units²
(b) 99 units²
(c) 40 units
(d) 20 units²

18. Using the number line below, which of the following expressions results in the smallest value?
(a) $-(abc)$

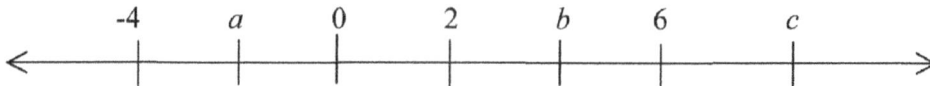

(b) $125a$
(c) $-10bc+a$
(d) $10(a+b+(-20))$

19. Use each of the values for the variables below and evaluate each expression below through substitution. Which expression will result in the greatest value?

$$a=-1$$
$$b=\tfrac{1}{4}$$
$$c=10$$
$$d=0.1$$

   (a) $abcd$
   (b) $100ab$
   (c) $a+b+c+d$
   (d) $2(3a+3b)$

20. Which of the following is *not* equal to 0.0053 km?
   (a) 5.3 m
   (b) 53 lbs.
   (c) 5,300 mm
   (d) 530 cm

21. Which of the following equations is parallel to the given line: $y=2x+8$ and passes through the point (1,4)?
   (a) $y=-\tfrac{1}{2}x+4.5$
   (b) $y=2x+2$
   (c) (0, 8)
   (d) $y=2x+4$

22. Determine the equation of the line that is perpendicular to $y=-\tfrac{1}{4}x+4$ and passes through the point (2,12).
   (a) $y=-\tfrac{1}{4}x+4$
   (b) $y=4x-2$
   (c) $y=-\tfrac{1}{4}x-2$
   (d) $y=4x+4$

23. If triangle ABC is reflected over the *y*-axis, what are the coordinates of the image point for C'?

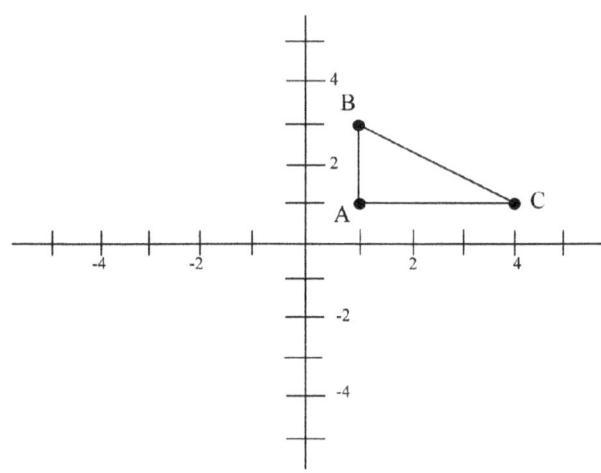

(a) (−4,1)
(b) (4,1)
(c) (4,−1)
(d) (−1,1)

24. A popcorn container, shaped like a cylinder, is 12 inches in diameter and 20 inches in height is packed in a box that is 14 inches long, 14 inches wide, and 22 inches high. What is the volume of the box that is not occupied by the container? (Round your answer to the nearest whole number.)
    (a) 2,050 in³
    (b) 4,312 in³
    (c) 6,574 in³
    (d) 4,072 in²

25. This number is between which two positive numbers?

    $$2\sqrt{13}$$

    (a) 4 and 5
    (b) 6 and 7
    (c) 7 and 8
    (d) 10 and 11

26. Amy earned $8 per hour for work study last semester. This semester she earned $8.50. Which describes the change in Amy's hourly earnings?
    (a) a decrease of 6.25%
    (b) an increase of 6.25%
    (c) a decrease of 8.25%
    (d) an increase of 8.25%

27. A hockey team played h games, losing six of them and winning the rest. What is an expression that shows a comparison of games won to games lost?
    (a) $\dfrac{h}{6}$
    (b) $\dfrac{6}{h}$
    (c) $\dfrac{6}{h-6}$
    (d) $\dfrac{h-6}{6}$

28. Given the right triangle BCD below, what is the measure of angle CDE?

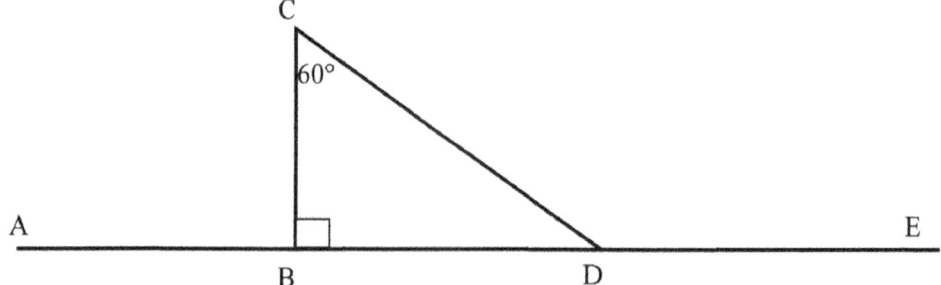

(a) 30°
(b) 150°
(c) 90°
(d) 45°

29. A Philadelphia airport shuttle service left the airport and traveled 21 miles directly south for a pick-up. It then took a sharp right-angle turn and drove straight for another 28 miles to reach its destination. How far is the shortest, most direct distance from the airport to the pick-up destination?
    (a) 35 miles
    (b) 49 miles
    (c) 98 miles
    (d) 20 miles

30. The U.S. Postal Service hires additional mail sorters each year for the holiday season. If six postal workers can sort 82 packages in 3 hours, how long will it take for three additional workers to sort the same number of packages?
    (a) 5 hours
    (b) 2 hours
    (c) 4.5 hours
    (d) 1 hour

31. Use the following mathematical description to answer the question that follows:

$$\frac{3}{5} < y < \frac{2}{3}$$

Which of the following values could be substituted for y to make the statement true?
    (a) $\frac{2}{3}$
    (b) $\frac{5}{8}$
    (c) $\frac{1}{4}$
    (d) $\frac{8}{9}$

32. What is the total amount paid for a loan of $1,050 at a simple interest rate of 3% for 4 years?
    (a) $126
    (b) $924
    (c) $1176
    (d) $1081.50

33. An assembly line of workers is able to assemble 780 cell phones in the first hour, 720 in the second hour, 660 cell phones in the third hour. If this pattern continues, how many cell phones will be assembled *after* the fifth hour?
    (a) 600
    (b) 540
    (c) 60
    (d) 3,300

34. Katie bought an equal number of $95, $75, and $50 tickets to an orchestra presentation of Beethoven's Moonlight Sonata. She spent $1,100 in all for the tickets. How many of each ticket did Katie buy?
    (a) 20
    (b) 5
    (c) 3
    (d) not enough information is provided to solve the problem
35. Which example below proves the following conjecture to be true: *Every even integer greater than 2 can be written as a sum of two prime numbers.*
    (a) 100=47+53
    (b) 46=13+33
    (c) 5=3+2
    (d) 52=25+27
36. The visual below describes the geometric relationship for the following procedure in algebra:

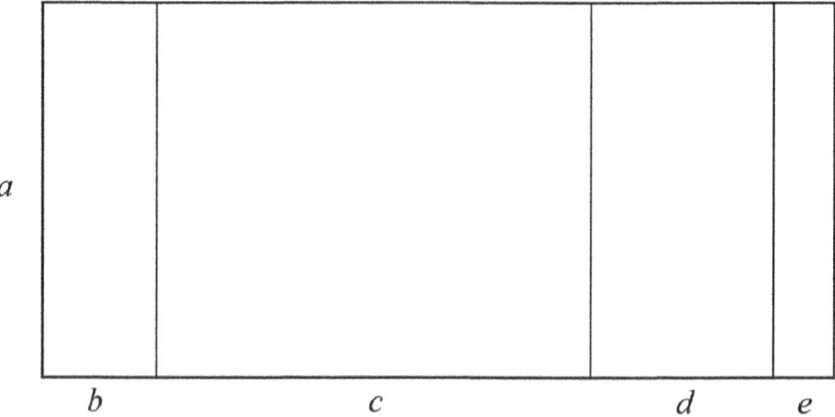

    (a) squaring a binomial
    (b) factoring a binomial
    (c) finding the greatest common factor
    (d) distributing a term across a polynomial

ANSWERS AND DETAILED EXPLANATIONS
FOR FULL PRACTICE TEST 3

1. Answer: (d) 42%
   Estimate your answer. The amount of viewers increased by 4.242 million people, which is not quite ½ (or a little less than 50%) of the original viewing audience of 10.1 million people. Answer choice (d) makes the most sense. Now do the calculations to check the estimate.
   First find the amount of increase by subtracting the usual audience from the finale audience: 14.342−10.1=4.242. Divide the amount of increase by the original amount: 4.242÷10.1=0.42. Convert the decimal 0.42 to a percent by multiplying it

by 100. The percent increase in the viewing audience of the finale is 42%. Answer choice (b) is the amount of increase and not the percent increase. Answer choices (a) and (c) don't make sense as answers if the audience went from 10.1 million to 14.342 million. These answers can be eliminated.

Revisit the original estimate with the calculated answer of 42% increase to see if the answer is reasonable. Yes, the calculations make sense.

Topic of Question 1: Percent of increase. More on percentages can be found in chapter 2.

2. Answer: (a) 30 minutes

   Convert the two hours into minutes: 2 hours = 120 minutes. Multiply the 120 minutes by the 3 available batting cages to get the total of minutes available for use in the cages: 120 minutes × 3 cages = 360 minutes of total batting-cage time reserved. Divide 360 by the number of players to find the equal number of minutes each player can use the cages: 360 minutes ÷ 12 players = 30 minutes practice time in the batting cages per player. Answer choices (b) and (c) are too low and would result in wasted time at the batting cages. Answer (d) is too much time, which would mean all players would not get to use the batting cages in the determined two hours.

   Topic of Question 2: Translating word problems as expressions. More on algebraic principles can be found in chapter 3.

3. Answer: (c) 3

   Multiply 6.25×10^3 to get 6,250. Multiply 3.1×10^4 to get 31,000. Multiply $6250 \times 31,000$ to get 193,750,000. 3 is in the millions place. Another strategy is to use the associative property by multiplying $6.25 \times 3.1 = 19.375$ and $10^3 \times 10^4 = 10^{3+4} = 10^7 = 10,000,000$. Multiply $19.375 \times 10,000,000$ to get 193,750,000. Answer choice (a) is not in the product and answers (b) and (d) do not represent the millions place.

   Topic of Question 3: Different representations of numbers. More on writing numbers in different forms can be found in chapter 2.

4. Answer: (a) 20"

   Since triangle ABC and triangle DEF are similar, their corresponding sides form proportions. It is important to first match up the corresponding sides for the triangles. Side AB corresponds to side DE. Side BC corresponds to side EF. Side CA corresponds to side FD. Therefore the following proportions are correct:

   $$\frac{2}{5} = \frac{8}{AB}$$

   To solve the proportion, cross-multiply and divide.
   Therefore, the length of side AB = 20".

   Topic of Question 4: Applying the relationship of corresponding sides of similar figures. More on this can be found in chapter 5.

5. Answer: (d) xy^2z

   x is a factor of both x^2 and x. y^2 is a factor of y^3 and y^2. z is a factor of z and z^2. Answer choice (a) is incorrect because z^2 is not a factor of the first term. Answer choice (b) is incorrect because x^2 is not a factor of the second term. Answer (c) is not correct because y^3 is not a factor of the second term.

   Topic of Question 5: Factoring. More on factoring can be found in chapter 3.

6. Answer: (d) $24b^2c+72b$

    The terms in the parentheses are unlike terms so they cannot be combined. Use the distributive property to multiply: $12b \times 2bc$ to get $24b^{1+1=2}c$ and $12b \times 6$ to get $72b$. The simplified expression equivalent to $12b(2bc+6)$ is $24b^2c+72b$. Answer choice (a) is incorrect because the two terms in the parentheses are unlike terms and cannot be combined. Answer choice (b) is wrong because the 6 was combined with the $24b^2c$ and that is incorrect because $24b^2c$ and 6 are unlike terms and cannot be combined. Answer choice (c) is wrong because $12b$ was not distributed over 6 to make $72b$.

    Topic of Question 6: Applying the distributive property. More on the distributive property can be found in chapter 3.

7. Answer: (d) 46°, 92°, 42°

    The sum of the angles of any triangle is 180°. This immediately eliminates answer choice (b), and answer choices (a) and (c) do not answer the question. The only option left to choose from that makes sense is answer choice (d). Now do the work to make sure it is reasonable.

    Create an algebraic equation to calculate the measure of the three angles. The first angle is unknown, which would represent the variable x. The second angle of a triangle is twice as long as the first, or $2x$. The third angle is 50° less than the second: $2x-50$. The sum of the three angle measurements is 180:

    $$x+2x+2x-50=180$$

    Collect like terms on the left side of the equation.

    $$5x-50=180$$

    Add 50 to both sides of the equation.

    $$5x=230$$

    Divide by 5.

    $$x=46$$

    Substitute 46 into the original equation to calculate the other two angle measurements.

    $$46+2(46)+[2(46)-50]=180$$

    The measurements of the three angles are 46°, 92°, and 42°.

    Topic of Question 7: Applying algebraic principles in word problems. More on algebraic principles in word problems can be found in chapter 3.

8. Answer: (b) 72.5

    The family already traveled 45 kilometers and still has 27,500 meters to travel. To calculate how many kilometers the trip is in all, first convert the meters to kilometers and find the sum. There are one thousand meters in one kilometer, so divide

27,500 by 1,000 to convert meters to kilometers. 27,500 meters = 27.5 kilometers. Add 45+27.5 to get the total distance of the trip in kilometers: 45+27.5=72.5 kilometers in all. Answer choice (a) is wrong because the measures taken are in different units. Answer choice (c) is the sum in meters and not kilometers. Answer choice (d) is 27,500 divided by 45, which doesn't answer the question and can be eliminated.
Topic of Question 8: Estimating and measuring in the metric system. More on the metric system can be found in chapter 5.

9. Answer: (a) 98
Make a reasonable estimate. The fifth score will need to be in the upper nineties if the student is going to earn an "A" because two of the four grades are lower than the necessary 90 average. The only choice that makes sense is (a).
Create an equation to solve the problem.

$$\frac{94+88+80+90+x}{5} = 90$$

Combine like terms on the left side of the equation.

$$\frac{352+x}{5} = 90$$

Multiply both sides of the equation by 5 to clear the fraction.

$$352+x=450$$

Subtract 352 from both sides of the equation to solve for x.

$$x=98$$

The other answer choices are too low for a student to achieve a 90 mean score.
Topic of Question 9: Measures of central tendency. More on mean, median, and mode can be found in chapter 6.

10. Answer: (b) $y=-x^2+5$
Because the graph includes a vertical parabola, the value of x must be squared. Therefore, answer choice (d) is eliminated. It is also true that because the parabola opens downward, the coefficient of the x^2 term must be negative, so that eliminates answer choice (a). The vertex of the parabola is the y-intercept, which is at 5. The only answer choice that is reasonable is choice (b) $y=-x^2+5$. Now, substitute (0,5) into the equation to check for accuracy.

$$y=-0^2+5$$
$$5=5$$

Answer choice (b) $y=-x^2+5$ is correct.
Topic of Question 10: Graphing parabolas. More on this can be found in chapter 4.

11. Answer: (c) 45
The Fundamental Counting Principle states that if there are m ways for one event to occur, and n ways for a second event to occur, then there are $m \times n$ ways for both to occur. There are three main entree options, five sides, and three dessert

options. Then, according to the Fundamental Counting Principle, there are 3 × 5 × 3=45 possible meals. Answer choices (a) and (b) are not reasonable answers and can be eliminated. Answer choice (d) does not make any sense to the problem as well and can be eliminated.
Topic of Question 11: The Fundamental Counting Principle. More on this can be found in chapter 6.

12. Answer: (a) $\frac{1}{2} \times \frac{1}{2}$

There are six possible face values for the cube and three of them are even numbers. Therefore, there are three possible outcomes out of the six that are even numbers (6, 12, 18), which is equivalent to $\frac{1}{2}$. There are three numbers between 1 and 10 (3, 6, 9), which results in three out of the six possibilities, also equaling $\frac{1}{2}$. In order to get the probability of both occurring, it is important to multiply the probability of each occurring.

Answer choice (c) is incorrect because that would be the result of one event occurring *or* the other occurring. Answer choice (b) is wrong because there are more than one even number and more than one number between 1 and 10 in the data. Answer choice (d) is not a reasonable answer and can be automatically eliminated.
Topic of Question 12: Probability. More on probability can be found in chapter 6.

13. Answer: (b) 336

A permutation is an arrangement or listing in which *order is important*. To find the permutation of *n* objects taken *r* at a time, use the following formula:

$$_nP_r = \frac{n!}{(n-r)!}$$

$$_8P_3 = \frac{8!}{(8-3)!}$$

$$\frac{8 \times 7 \times 6 \times 5!}{5!}$$

Simplify the fraction to get 8 × 7 × 6=336 different ways to arrange which dog would be awarded the gold, silver, or bronze medal. Answer choice (a) is incorrect because it just gives the number of dogs in the contest. Answer choice (c) is not a reasonable answer and can be eliminated as an option. Answer choice (d) is how many combinations are possible. Combinations do not consider order, and when awarding ribbons in a competition, order is very important.
Topic of Question 13: Permutation. More on permutations and combinations can be found in chapter 6.

14. Answer: (a) (−2)(−1)

A counterexample will make the statement untrue. Try each answer choice. Answer choices (b), (c), and (d) all result in a product that is smaller than the factor multiplied by (−1), making the statement true. Answer choice (a) is (−2)(−1)=2. The product is greater than the factor multiplied by (−1). Answer choice (a) is a counterexample to the above statement.
Topic of Question 14: Different representations of numbers. More on writing numbers in different forms can be found in chapter 2.

15. Answer: (c) Most fans enjoy watching football and hockey, but not baseball.

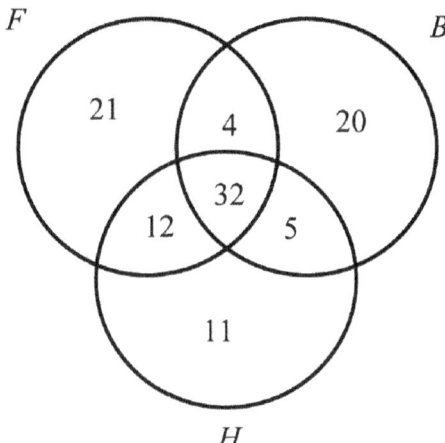

The Venn Diagram above shows that most sports fans watch more than one sport; football is the favorite sport to watch, while hockey is the least-favorite sport to watch. Answer choice (c) most fans enjoy watching football and hockey, but not baseball, is untrue.

Topic of Question 15: Venn Diagrams. More on Venn Diagrams can be found in chapter 6.

16. Answer: (d) box-and-whisker plot
A box-and-whisker plot is a graph that displays a data set using quartiles and the median. It takes raw data and illustrates the spread and distribution. This would be most appropriate to highlight the reading scores of the thirty students. Answer choice (a) is incorrect because a line graph is a graph that shows changes in data over time. Answer choice (b) is incorrect because a bar graph is a graph used to show relationships or comparisons between categories. Answer choice (c) is wrong because a circle graph is best used to compare part to the whole and parts to other parts.

Topic of Question 16: Choosing appropriate ways to show data. More on appropriateness of graphs can be found in chapter 6.

17. Answer: (b) 99 units^2
The formula for finding area of a rectangle is $A=lw$. The first thing to do is to find the measurement of the length and width of the rectangle. The perimeter of a rectangle is 40 units with a length of $(2x-1)$ and the width is $(x+6)$. Create an equation to solve for x.
Use half the perimeter since only the length and width are needed.

$$(2x-1)+(x+6)=20$$

Collect like terms.

$$3x+5=20$$

Subtract 5 on both sides of the equation.

$$3x=15$$

Divide both sides of the equation by 3.

$$x=5$$

Substitute 5 in for x in each of the following expressions:

$$2x-1=2(5)-1=9$$

The length is nine units.

$$x+6=5+6=11$$

The width is eleven units.

$$A=lw=9 \times 11=99 \text{ units}^2$$

Answer choice (a) is not a reasonable answer and can be automatically eliminated. Answer choice (c) is the given perimeter and is not in square units. It can also be eliminated. Answer choice (d) is wrong because while identifying the length and width, the area calculations are incorrect. Answer choice (b) is the only feasible answer.

Topic of Question 17: Applying formulas to calculate and determine lengths. More on using formulas to calculate lengths can be found in chapter 5.

18. Answer (c) $-10bc+a$

    First identify the interval (space between the tick marks). The interval is 2.
    Find the value of each of the variables:

    $$a=-2$$
    $$b=4$$
    $$c=8$$

    Substitute the values of each of the variables and evaluate each expression.
    (a) $-(abc)=-(-2 \times 4 \times 8)=-(-64)=64$
    (b) $125a=125 \times -2=-250$
    (c) $-10bc+a=-10 \times 4 \times 8+(-2)=-320+(-2)=-322$
    (d) $10(a+b+c)=10(-2+4+8)=10(-18)=-180$
    Answer choice (c) -322 is the least value.

    Topic of Question 18: Relative magnitude of real numbers. More on real numbers can be found in chapter 2.

19. Answer: (c) $a+b+c+d$

    Use substitution to evaluate each expression. After simplifying each expression through substitution, $a+b+c+d$ results in the greatest number.
    Answer choice (a) $abcd=(-1)\times\frac{1}{4}\times 10\times 0.1=-\frac{1}{4}$
    Answer choice (b) $100ab=100\times(-1)\times\frac{1}{4}=-25$
    Answer choice (c) $a+b+c+d=(-1)+\frac{1}{4}+10+0.1=9.35$—the greatest value
    Answer choice (d) $2(3a+3b)=2(3\times(-1)+3\times\frac{1}{4})=2(-3+0.75)=2\times-2.25=-4.5$

    Topic of Question 19: Substitution and evaluating expressions. More on expressions can be found in chapter 3.

20. Answer: (b) 53 lbs
Answer choice (b) clearly is not equal to 0.0053 kilometers. It is not in the metric system. Even within the U.S. Customary System of measurement, pounds measure weight and not length. Answer choices (a), (c), and (d) are all equivalent: 0.0053 kilometers = 5.3 meters = 530 centimeters = 5,300 millimeters.
Topic of Question 20: Estimating and measuring in the metric system. More on the metric system can be found in chapter 5.

21. Answer: (b) $y=2x+2$
Two parallel lines share the same slope. The slope is the *x*-coefficient, m, when a linear equation is written in slope-intercept form: $y=mx+b$. A line parallel to $y=2x+8$ must have 2 as the slope. Therefore, that automatically eliminates choices (a) and (c). Choice (a) $y=-\frac{1}{2}x+4.5$ is perpendicular to the given line because the slopes are negative reciprocals of each other. Choice (c) is the *y*-intercept (0,8) of the given line. Both choices do not answer the question at hand.
Answer choices (b) and (d) are both parallel to the given line. Substitute the values of the given point to see which line fits both conditions. Begin with answer choice (b).

$$y=2x+2$$
$$4=2(1)+2$$
$$4=4$$

Point (1,4) is a solution to the equation y=2x+2. Thus, the line passes through point (1,4).
The point (1,4) is not a solution to answer choice (d) $y=2x+4$.

$$4=2(1)+4$$
$$4\neq 6$$

Topic of Question 21: Equations of parallel and perpendicular lines. More on relating linear equations can be found in chapter 4.

22. Answer: (d) $y=4x+4$
Perpendicular lines have negative reciprocal slopes. The slope is the *x*-coefficient, m, when a linear equation is written in slope-intercept form: $y=mx+b$. Therefore a line perpendicular to $y=-\frac{1}{4}x+4$ would have to have a slope of 4. Answer choices (a) $y=-\frac{1}{4}x+4$ and (c) $y=-\frac{1}{4}x-2$ can be eliminated. Both choices are parallel to the given line because they have the same slope.
Substitute the given point (2,12) to see if it is a solution to answer choice (b):

$$y=4x-2$$
$$12=4(2)-2$$
$$12\neq 6$$

Answer choice (b) does not pass through the point (2,12).
Answer choice (d) should be the answer.

$$y=4x+4$$
$$12=4(2)+4$$
$$12=12$$

Answer choice (d) is perpendicular to the given line and passes through point (2, 12).
Topic of Question 22: Equations of parallel and perpendicular lines. More on relating linear equations can be found in chapter 4.

23. Answer: (a) (−4,1)

    Triangle ABC reflected over the *y*-axis would result in an image with the same *y*-coordinate for each image point and an opposite *x*-coordinate for each image point. Therefore, the image of point C (4,1) reflected over the *y*-axis would be C' (−4,1).

    Answer choice (b) (4,1) are the coordinates of the point C. It does not show the reflection and can be eliminated as a possible answer. Answer choice (c) (4,−1) would be the coordinates of C' if it were reflected over the *x*-axis. Answer choice (d) (−1,1) are the coordinates for point A reflected over the *y*-axis. It does not answer the question. The only answer choice that is reasonable is (a) (−4,1).

    Topic of Question 23: Graphing transformations. More on transformations can be found in chapter 5.

24. Answer: (a) 2,050 in³

    In order to find how much space is not occupied by the popcorn container, use the following formula:

    $$V_{box} - V_{container} = V_{\text{space not occupied by the container}}$$

    Calculate the volume of the box.

    $$V = lwh$$
    $$V = 14 \times 14 \times 22$$
    $$V = 4312 \text{ in}^3$$

    Calculate the volume of the popcorn container.

    $$V = \pi r^2 h$$

    (The diameter was given. Divide the diameter by 2 to get the radius.)

    $$V = \pi \times 6^2 \times 20$$
    $$V = 2261.946711$$

    Round to the nearest whole number.

    $$V \approx 2262 \text{ in}^3$$

    Go back to the original formula:

    $$V_{box} - V_{container} = V_{\text{space not occupied by the container}}$$

    $$4312 - 2262 = 2050 \text{ in}^3$$

Answer choice (b) is incorrect because it is the volume of the box and not the difference between the box and the popcorn. Answer choice (c) doesn't make sense because the space not occupied by the popcorn cannot be bigger than the box itself. Answer choice (d) can be eliminated because it is not in cubic units and volume is measured in cubic units.

Topic of Question 24: Applying formulas to calculate and determine volume. More on using formulas to calculate volume can be found in chapter 5.

25. Answer: (c) 7 and 8
Estimate by rounding up the square root of 13 to a perfect square:
The next perfect square is

$$\sqrt{16}$$

Calculate the square root of 16, which equals 4, and then multiply that by 2. The answer will be no greater than 8, but should be close to 8. Therefore, answer choices (a) and (d) can be automatically eliminated.
Simplify $2\sqrt{13}$
Calculate the square root of 13 first and then multiply that answer by 2.

$$2\sqrt{13} \approx 7.211102551$$

This number is between the two positive numbers 7 and 8.
Topic of Question 25: Display understanding and fluency in computation. More on computation of real numbers can be found in chapter 2.

26. Answer (b) an increase of 6.25%
First estimate; Amy got an increase in her salary. This eliminates choices (a) and (c). Amy got a $0.50 per-hour raise. An increase of 12% (d) seems too high of an increase percentage for a $0.50 per-hour raise. Answer choice (b) is the most reasonable answer.
Check your estimate by using the following formula for finding percent of increase or decrease:

$$\frac{\text{New Value} - \text{Original Value}}{\text{Original Value}} \times 100$$

$$\frac{(8.50-8.00)}{8.00} \times 100 = 6.25\% \text{ increase}$$

Topic of Question 26: Percent of increase. More on percentages can be found in chapter 2.

27. Answer: (d) $\frac{h-6}{6}$
The hockey team lost six games and won all the others ($h-6$). The comparison must be games won to games lost. Therefore, answer choices (b) and (c) can be eliminated because the ratio comparison is games lost to games played or games won. Answer choice (a) is also wrong because the team did not win h games, but ($h-6$) games. The only reasonable answer is answer choice (d).
Topic of Question 27: Applying algebraic principles in word problems involving ratios. More on algebraic principles involving ratios can be found in chapter 3.

28. Answer: (b) 150°

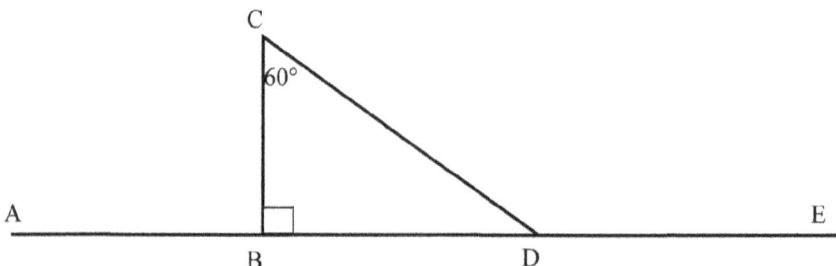

Answer choice (b) 150° is clearly the only answer that makes sense since angle CDE is an obtuse angle. Now calculate the angle measurement to make sure it is right. Since triangle BCD is a right triangle and the sum of the angles of any triangle equals 180°, angle BDC = 30°. Angles BDC and CDE form a linear pair, so the measurement of angle CDE must be 150° since the measurement of angle BDC is 30°.

Topic of Question 28: Attributes of triangles and angle relationships. More on triangles can be found in chapter 5.

29. Answer: (a) 35 miles

The shuttle service is traveling 21 miles directly south, and then right 28 straightforward miles. Draw a diagram of the directions. It forms the two legs of a right triangle. In order to calculate how far the shortest distance is from the airport to the pick-up destination, use the Pythagorean Theorem in order to find out the hypotenuse.

$$a^2+b^2=c^2$$
$$21^2+28^2=c^2$$
$$1225=c^2$$

The square root of 1,225 is 35. The most direct route to the airport is 35 miles.
Answer choice (d) 20 miles, can be eliminated because 20 miles is shorter than the 21 miles the shuttle took south. Answer choice (b) is wrong because it is the distance from the airport to the destination using the original route. This is not the shortest distance to the airport. Answer choice (c) can be eliminated because it is not the distance from the airport to the destination. It is the total distance the shuttle traveled during the trip.

Topic of Question 29: The Pythagorean Theorem. More on estimating and calculating measurements using the Pythagorean Theorem can be found in chapter 5.

30. Answer: (b) 2 hours

First estimate the answer. Because there are more sorters hired, the time to sort the same amount of packages will be less. Therefore, answer choices (a) and (c) can be eliminated. Answer choice (c) is a common error. It is the result of setting this equation up as a direct proportion. Using estimation first will help in eliminating this answer choice.

This is an inverse proportion showing that one quantity increases as a related quantity decreases. To calculate the time with three additional workers, set up the following equation:

fewer workers × time with those workers
= more workers × x (time with more workers)

$$6 \times 3 = 9x$$
$$18 = 9x$$
$$x = 2 \text{ hours}$$

Replace two hours into the original equation to check for accuracy. Revisit the original estimate. Answer choice (b) makes sense.

Topic of Question 30: Applying algebraic principles in word problems. More on algebraic principles in word problems can be found in chapter 3.

31. Answer: (b) $\frac{5}{8}$

    It is easiest to convert each of the fractions to decimals:

    $$\frac{3}{5} = 0.6$$
    $$\frac{2}{3} = 0.\overline{6}$$

    Answer choice (a) $\frac{2}{3}$ is not a reasonable answer because it cannot be less than itself. It can be eliminated.

    Answer choice (b) $\frac{5}{8} = 0.625$, which is between 0.6 and $0.\overline{6}$.

    Answer choice (c) $\frac{1}{4} = 0.25$, which is less than 0.6.

    Answer choice (d) $\frac{8}{9} = 0.\overline{8}$, which is greater than $0.\overline{6}$.

    Topic of Question 31: Different representations of numbers. More on writing numbers in different forms can be found in chapter 2.

32. Answer: (c) $1,176

    First estimate the answer. Since it is a loan charging interest, the amount paid back will be more than the amount borrowed. That eliminates answer choices (a) and (b). This will avoid unreasonable answers. Answer choice (a) $126 is a common error in this type of problem. It is the interest paid, but not the total amount of the loan paid back. Answer choice (b) is another common error. It is the interest subtracted from the loan amount. With a loan, the interest will be added to the principal loan. It is not a discount problem. Using estimating first will help avoid common errors. Use the provided formula for simple interest:

    $$I = Prt$$
    $$I = 1050 \times 0.03 \times 4$$

    $I = \$126$ interest paid over the four years at 3% interest

    To calculate the total amount paid, add the interest to the principal loan.

    $$\$1050 + \$126 = \$1176$$

    Topic of Question 32: Simple interest. More on simple interest can be found in chapter 2.

33. Answer: (d) 3,300
First discover the sequence. A sequence is an ordered set of items or numbers that follows a pattern. In this case, there are 60 fewer cell phones being assembled each hour. Recognize that the question is asking for the total number of cell phones assembled after five hours.

first hour	second hour	third hour	fourth hour	fifth hour	total
780	720	660	600	540	3,300

Answer choice (a) is incorrect because it is the number of cell phones assembled in the fourth hour. Answer choice (b) is incorrect because it is the number of cell phones assembled in the fifth hour. Answer choice (c) simply identifies the pattern of sixty fewer cell phones assembled each hour.
Topic of Question 33: Identifying patterns. More on patterns can be found in chapter 3.

34. Answer: (b) 5
Set up an equation to calculate the number of each ticket Katie bought. One variable can be used because she bought an equal amount of each ticket.

$$95x+75x+50x=1100$$

Combine like terms: $220x=1100$.
Divide by 220 to get the number of each ticket Katie purchased: $x=5$ tickets each.
Place the answer back into the original equation:

$$95(5)+75(5)+50(5)=1100$$
$$475+375+250=1100$$
$$1100=1100$$

Answer choice (d) is not correct because there is sufficient information to solve the problem. Answer choice (a) doesn't make sense because if Katie bought 20 tickets at $95 each ticket, she wouldn't have enough money to pay for this purchase alone since $20 \times 95=1900$.
Topic of Question 34: Applying algebraic principles in word problems. More on algebraic principles in word problems can be found in chapter 3.

35. Answer: (a) 100=47+53
Which example below proves the following conjecture to be true? *Every even integer greater than 2 can be written as a sum of two prime numbers.*
Answer choice (a) shows the conjecture to be true. 100 is an even integer greater than 2 and is written as a sum of two prime numbers. Answer choice (b) is an even integer greater than 2, but although 13 is a prime number, 33 is not prime. The sum of the addends in answer choice (c) is an odd number, so it is not an example that proves the above conjecture to be true. Answer choice (d) is not an example that proves the conjecture to be true because the addend 25 is not a prime number.
Topic of Question 35: Demonstrate understanding of real numbers. More on real numbers can be found in chapter 2.

36. Answer: (d) distributing a term across a polynomial
    Identify that the figure represents a length and width that includes one term a being distributed over four terms (b, c, d, and e). Therefore, geometrically a is being distributed across the polynomial: $a(b+c+d+e)$. Answer choice (d) distributing a term across a polynomial is the correct choice.
    Topic of Question 36: Distributive property. More on displaying distribution can be found in chapter 3.

Index

active reading, 3
acute angles, 99, *99*
addition: one-variable equations solved by, 33–34; systems of linear equations solved by, 62–65; words and phrases for, 37
algebraic expressions: order of operations for, 41; from word problems, 37–41
alternate angles, *101*, 101–2
angles: acute, 99, *99*; alternate, *101*, 101–2; classification of, *101*, 101–2; complementary, 100; definition of, 99; exterior, *101*, 101–2; interior, *101*, 101–2; obtuse, 99 *99*; right, 99, *99*; of right triangles, 87–88, *88*; straight, 99, *99*; supplementary, 100; of triangles, 86–87
answers, on tests: checking, 3; eliminating, 6; estimating, 6; types of, 5
area: of circles, 84–86; of geometric figures, 84; of irregular figures, 94–97; of parallelograms, 92; of rectangles, 92; of shaded regions, 94–97; of squares, 93
attitude, in test-taking, 5, 6

bar graphs, 116, *116*
binomials, multiplication of, 42–43
box-and-whisker plots, 117–18
brain breaks, during testing, 7
breaks, during testing, 7

capacity: metric units of, 81–84; U.S. customary units of, 79–81
Cartesian coordinate system, 102–4
center of a circle, 84, *85*

centi-, 81
central tendency, measures of, 109–12
chord, 84, *85*
circle, area and circumference of, 84–86, *85*
circle graphs, *113*, 113–14
circumference of a circle, 84–86, *85*
coefficients, in one-variable equations, 34
combinations, 132–35
common denominator, in solving equations, 35–36
common sense, in eliminating answers, 6
complementary angles, 100
composite numbers, 9
compound interest, 21–23
compound probabilities, 126, *126*
comprehension, monitoring and repairing, 2
conditional equations, 33
conditional probability, 127–28
congruent triangles, 90, *90*
consecutive integers, 11
coordinate geometry, 102–4
counting numbers, 9
counting principles, 131–34
cross-multiplication, 49–50

data: analyzing, 118–21; bar graphs for, 116, *116*; circle graphs for, *113*, 113–14; histograms for, *114*, 114–16, *116*; inferences from, 118–21; tables for, *112*, 112–13
days, measuring with, 79–81
deka-, 81
deci-, 81

decimals, 12–13; equations with, 36; finite, 9; fractions and, 15–16; infinite, 9; percents and, 15–16; practice test questions on, 14–15; probabilities as, 122
dependent probability, 127
Descartes, Rene, 102
diameter, 84, *85*
distance between two points, 103
distributive property: in binomial multiplication, 42–43; in solving equations, 34
division: of exponential terms, 25; words and phrases for, 37
dot patterns, 28–29
double star, 2–3

elimination, solving systems of linear equations by, 62–64
equality, in generating algebraic expressions, 37–41
equations: combining like terms in, 42; definition of, 33; in generating algebraic expressions, 37–41; one-variable, 31–36; order of operations for, 41; for parabola, 73; simplifying, 41
equiangular triangles, 87, *87*
equilateral triangles, 87, *87*
exponent(s): in exponential notation, 24; negative, 25; practice test questions on, 25–26
exponential notation, 24–26
expressions, factoring, 43–47
exterior angles, *101*, 101–2

factor boxes, 45–46
factoring, of expressions, 43–47
finite decimals, 9
fix-up tools, for comprehension, 2
fluid ounce, 79–81
FOIL (first, inner, outer, last), 42
foot (feet), 79–81
formulas: for area of a circle, 84; for area of quadrilaterals, 92; for circumference, 84; for compound interest, 21; for distance between two points, 104; for midpoint of a line, 103–4; for perimeter of quadrilaterals, 92; for probability, 122; quadratic, 46–47; for simple interest, 20; for slope, 53; for surface area of a rectangular prism, 95–96
fractions: decimals and, 15–16; equations with, 35–36; percents and, 15–16; probabilities as, 122; ratios as, 48
frequency distributions, 118, *119*
functions, nonlinear, 73–77
Fundamental Counting Principle, 130–31
future value, in compound interest, 21
gallon, 79–81

geometric figures: area of, 83; irregular, 94–97; perimeter of, 83; surface area of, 84; volume of, 84
gram, 81–84
graphing: of lines, 53–60; of parabolas, 73–75; of parallel lines, 59–62; of perpendicular lines, 59–62;
graphs, of data: bar graphs, 116, *116*; circle graphs, *113*, 113–14; definition of, 119; histograms, *114*, 114–16, *116*; tables, *112*, 112–13

hecto-, 81
histograms, *114*, 114–16
horizontal lines, intercepts of, 56
hours, 79–81
hypotenuse, of right triangles, 87–88, *88*

identities, 33
inch, 79–81
inclusive sets, 11
inconsistent equations, 33
inequalities: one-variable, solving algebraically, 32–37. *See also* linear inequalities
inequality signs, in linear inequalities, 68–69
infinite decimals, 9
inspection, graphing lines from, 56, 61
integers: consecutive, 11; definition of, 9; multiplication of, 11; sets of, 11
intercepts, graphing lines from, 54–56
interest: compound, 21–22; practice test questions on, 23–25; simple, 20–21
interior angles, *101*, 101–2
intersections of sets, 123, *123*
irrational numbers, 9
irregular figures, area and perimeter of, 94–96
isosceles triangles, 87, *87*

kilo-, 81
King Henry Died Drinking Chocolate Milk, 82

legs, of right triangles, 87–88, *88*
length: metric units of, 81–84; U.S. customary units of, 79–81
like terms, combining, 42
line(s): graphing, 53–60; midpoint of, 103–4; slope of, 53; transversal, *101,* 101–2. *See also* parallel lines; perpendicular lines
linear equations: graphing, 53–60;
linear inequalities: graphing, 67–69; inequality signs in, 68–69; solving, 68; writing, 68
liter, 81–84

mean, 109
measurement: of geometric figures, 84; metric system, 81–84; unit conversion, 84; U.S. customary system, 79–81
measures of central tendency, 109–12
median, 109–10
mental stamina, in test-taking, 7
meter, 81–84
metric system of measurement, 81–84
midline of the body, crossing, 7
midpoint of a line, 103–4
mile, 79–81
milli-, 81
minutes, 79–81
mode, 109–10
multiplication: of binomials, 42–43; of exponential terms, 25; of integers, 11; words and phrases for, 37; zero product property for, 46–47
Multiplication Principle, 130–31

natural numbers, 9
negative exponents, 25
negative slope, 53
nonlinear functions, 73–77
note taking, active, 3
number systems, 9–12, *10*

obtuse angles, 99, *99*
one-variable equations, solving algebraically, 31–35
one-variable inequalities, solving algebraically, 31–35

operations patterns, 29–30
ordered pairs: in Cartesian coordinate system, 102; graphing lines from, 55. *See also* points
order of operations, 41
origin, in Cartesian coordinate system, 102
ounce, 79–81

pacing, in test-taking, 5, 6
paper cutting patterns, 27
parabolas, 73–77
parallel lines: definition of, 59; graphing, 59–62; transversal through, *101,* 101–2
parallelograms, properties of, 92, *92*
patterns, analyzing and extending, 27–31
percent(s), 15–20; decimals and, 15–16; finding, 17–18; fractions and, 15–16; practice test questions on, 18–20
percent decrease, 18
percentiles, 119
percent increase, 18
perimeter: of geometric figures, 83; of irregular figures, 94–97; of parallelograms, 92; of rectangles, 92; of shaded regions, 94–97; of squares, 93; of triangles, 90
permutations, 132, 133–35
perpendicular lines: definition of, 60; graphing, 59–61; intersecting angles of, 100
pint, 79–81
place value, in decimal system, *12*
Please Excuse My Dear Aunt Sally (PEMDAS), 41
points: distance between, 104; transformations of, 104–9. *See also* ordered pairs
polynomials, factoring, 43–48
positive messages, in test-taking, 6
positive slope, 53
pound, 79–81
powers of ten, in metric system, 81
practice test(s), 3, 4, 135–58, 159–80, 181–203
practice test questions: on algebraic expressions, 39–40; on compound interest, 21–23; on converting numbers, 18–20; on counting principles, 133–34; on data analysis, 119–22; on data

display, 117–18; on decimals, 14–15; on exponents, 25–26; on measurement, 80–81; on measures of central tendency, 111–12; on number systems, 11–12; on patterns, 29–32; on percents, 18–20; on polynomial factorization, 47–48; on probability, 128–30; on proportions, 51–52; on ratios, 51–52; on simple interest, 23–24; on spread, 111–12

Pre-Service Academic Performance Assessment, 1; format of, 2; objectives of, familiarity with, 1–2; practice tests, 3, 4, 135–58, 159–80, 181–203; preparation for, 1–4; strategies for, 4–7. *See also* practice test questions

prime numbers, 9

principal: compound interest on, 21; simple interest on, 20

probability: compound, 126, *126;* conditional, 127–28; dependent, 127; formula for, 122; independent, 127–28; practice test questions on, 128–30; simple, 122–26; Venn diagrams for, *124,* 124–26, *125*

product, 11

proportions: definition of, 49; uses of, 50; word problems involving, 48–52

Pythagorean theorem, 87–89

quadrants, in Cartesian coordinate system, 102–3

quadratic formula, for factoring polynomials, 46–47

quadratic trinomials, factoring, 43

quadrilaterals, measuring, 94–97, *95*

quart, 79–81

questions, difficulty answering, 5

radius, 84, *85*

rate: for compound interest, 21; for simple interest, 20

ratio: definition of, 48; word problems involving, 48–51

rational numbers, 9

real numbers, 9

real-world situations: for linear equations, 57–58; for nonlinear functions, 73–74

rectangles, properties of, 92, *92*

rectangular prisms, volume and surface area of, 95–96

reflection, of a point, 104

registration, for the test, 1

rereading, in comprehension, 3

right angles, 99, *99*

right triangles, 88–91, *89;* the 30-60-90 triangle, *89,* 89–90; the 45-45-90 triangle, *88,* 89–90

rotation, of a point, 106–7

rounding, of decimals, 12–13

same-base product rule, 25

same-base quotient rule, 25

scalene triangles, 86, *87*

scientific notation, 13–15

seconds, measuring with, 79–81

sequences, analyzing and extending, 28

sets: of integers, 11; intersections of, 123, *123;* in probabilities, 125; union of, 123, *123*

shaded regions, area and perimeter of, 94–97

similar triangles, 90, *90*

simple interest, 20–21, 22–24

simple probability, 122–26

simplification, of equations, 41

slope of a line: definition of, 53; of parallel lines, 60; of perpendicular lines, 60

spheres, volume and surface area of, 97–98

spread, 109–12

squares, properties of, 93

statistics: analyzing data, 118–21; displaying data, 112–118; measures of central tendency, 109–12

stem-and-leaf plots, 115–17, *115, 116*

straight angles, 100, *100*

stretching, 7

substitution, solving systems of linear equations by, 66–67

subtraction: one-variable equations solved by, 33–34; words and phrases for, 37

supplementary angles, 100

surface area: of geometric figures, 83; of rectangular prism, 95–96; of a sphere, 97–98

systems of linear equations: elimination method for, 62–66; graphing of, 63–64;

solving, 63–69; substitution method for, 66–67

tables, for data, *112*, 112–13
test format, 2
test objectives, familiarity with, 1–2
test preparation, 1–4
test site, 4
test-taking strategies, 4–7
time: in compound interest, 21; in simple interest, 20; units of, 79–81
time management, in test-taking, 5, 6
ton, 79–81
transformations, of points, 106–8
translation, of a point, 106–7
transversal, *100*, 101–2
tree diagrams, for counting principles, 131–32, *132*
triangles: angles of, 86–87; congruent, 90, *90*; equiangular, 87, *87*; equilateral, 87, *87*; isosceles, 87, *87*; perimeter of, 90; right, 87–90, *88*; scalene, 86, *87*; sides of, 88–90; similar, 90, *90*
trinomials, factoring, 43–48

undefined slope, 53
unions of sets, 123, *123*
U.S. customary system of measurement, 79–81

variable: definition of, 31; eliminating, in solving systems of linear equations, 64–67; in generating algebraic expressions, 37–41
Venn diagrams, for probabilities, *124*, 124–26, *125*
vertical lines, intercepts of, 54–55
volume: of geometric figures, 84; of a rectangular prism, 95–96; of a sphere, 97–99; units of. *See* capacity

weight: metric units of, 81–84; U.S. customary units of, 79–81
whole numbers, 9
word problems: algebraic expressions from, 37–41; with proportions, 48–52; with ratios, 48–52

x-axis, 102
x-intercept, graphing lines from, 54–56

yard, 79–81
y-axis, 102
years, measuring with, 79–81
y-intercept, graphing lines from, 54–56

zero power, 24
zero product property of multiplication, 46–47
zero slope, 53

DATE DUE

PROPERTY OF MURRELL LIBRARY
MISSOURI VALLEY COLLEGE
MARSHALL, MO 65340